INTRODUCTION TO
SUPERCONDUCTIVITY
AND HIGH-T$_c$ MATERIALS

Michel Cyrot
Université J. Fourier & CNRS, Grenoble

Davor Pavuna
Ecole Polytechnique Federale de Lausanne

INTRODUCTION TO SUPERCONDUCTIVITY AND HIGH-T$_c$ MATERIALS

World Scientific
Singapore • New Jersey • London • Hong Kong

Published by

World Scientific Publishing Co. Pte. Ltd.

P O Box 128, Farrer Road, Singapore 9128

USA office: Suite 1B, 1060 Main Street, River Edge, NJ 07661

UK office: 57 Shelton Street, Covent Garden, London WC2H 9HE

First published 1992
First reprint 1995

Library of Congress Cataloging-in-Publication Data
Cyrot, M.
    Introduction to superconductivity and high-T_c materials / Michel
Cyrot, Davor Pavuna.
        p.   cm.
    Includes bibliographical references and index.
    ISBN 9810201435 -- ISBN 9810201443 (pbk)
    1. Superconductivity.   2. High-temperature superconductors.
I. Pavuna, Davor.   II. Title.
QC611.92.C97   1992
537.6'23 -- dc20   92-14874
                                                             CIP

Printed in Singapore.

FOREWORD

Superconductivity — like turbulence — stood as a major scientific mystery for a large part of this century. After fifty years of search the key concept (the Cooper pairs) emerged — and a number of earlier visions, by London, Landau and Ginzburg, etc. ..., became logically obvious.

On the other hand, the many-body theory of superconductors, with all its sophistication, appeared as such a perfect, complete object that it discouraged new thought. I, for one, would certainly not have recommended a high level of funding in superconducting materials ten years ago. Since the moment where mixed valency copper systems were found, the community of theorists became a tower of Babel–with different languages, different mystics, and different high priests.

It is bold to write a book on superconductivity in such a state of flux. But there is a need for a simple presentation of facts and rules. Graduate students who want to learn about high-T_c materials are exposed to this plague of our time: the <u>multi-author-book</u>, summarizing some recent conference, and edited by some enthusiast. The existence of these books is artificially imposed on us by the funding agencies. But they are essentially useless for a beginner — because of their patchwork structure, and also because the papers have no pedagogical motivation.

Under these unhappy conditions, an introductory text, with a unified, balanced point of view, is of considerable value. This is what Cyrot and Pavuna have produced. Their book still requires a significant effort for a genuine beginner, but it can be studied step by step. It sets up delicate compromises between the opposite dangers of dogmatism and oversimplification. I know that it will be of great help, and I wish for it the best of success.

P. G. de Gennes
Paris, August 7, 1991

v

PREFACE

This textbook is an introductory course that we address to students and researchers in all branches of science and engineering with a possible exception of theoretical physicists who may require a more mathematical approach.

Since the beginning of the 'superconductivity revolution' (in 1986-87) we felt that there is a need for a simple introductory textbook on superconductivity. We were asked to give introductory courses so this textbook naturally emerged from graduate lectures given at the Université Joseph Fourier in Grenoble and in CERN-Geneva (by MC) and from the final year undergraduate lectures given at the Swiss Federal Institute of Technology in Lausanne (by DP).

Most of the materials have been arranged in order of increasing difficulty. The student can *gradually* learn the key concepts which we deliberately repeat several times at even higher level. All chapters are reasonably self-contained and an experienced scientist can study them separately. The summary at the end of each chapter gives an overview of the most important concepts and useful formulae.

In Chapter 1 we introduce elementary notions on superconductivity and survey the main conventional superconducting materials. Chapter 2 gives a summary of the characteristic properties of superconductors. In Chapter 3, following a simple London model, we give a rather extensive discussion of the phenomenological Ginzburg-Landau theory which permits many practical calculations and predictions. Chapter 4 is intended for those who are involved in high critical current research, while Chapter 6 is an introduction to the physics of Josephson electronics. In Chapter 5 we give the essence of the microscopic BCS theory and some useful formulae; readers interested in more advanced approach are encouraged to study one of the books listed at the end of that Chapter. Finally, in Chapter 7 we introduce the characteristic properties of high-T_c oxides while in Chapter 8 we summarize the most relevant notions on the applications and technology of superconductors.

Despite all our efforts to avoid mistakes we are aware that many corrections and improvements remain to be made. We shall be grateful to any reader who would kindly suggest further improvements of the present text.

We are very grateful to all our teachers, colleagues, coworkers, students and friends in Grenoble, in Lausanne and elsewhere who have directly and indirectly contributed to our manuscript.

In particular we wish to thank P. G. de Gennes, who himself wrote a classic text on superconductivity in 1966, and who has kindly written the foreword. We thank O. Symko for numerous improvements in our preview-manuscript and N. W. Ashcroft, B. Chakraverty, G. Deutscher, A. Gilabert, A. Hebard, Ph. A. Martin, and F. K. Reinhart for many corrections and useful suggestions. We gratefully acknowledge all contributors and respective publishers who allowed us to adopt their illustrations and tables; their names and references are given at the end of the book (pages 235–237). We thank Prof. Y. K. Lim and Ms. U. Oesterle (and her mum) for editing our English.

We are grateful to our editor Ms. Kim Tan for her guidance, Prof. K. K. Phua and Ms. Faridah Shahab for their friendly encouragement.

Finally we thank the French and the Swiss scientific communities.

Last but not least we thank our families, Françoise and Sylvie, Christophe, Jean-Luc, Laurent and Marko.

There are many others whom we unintentionally forgot in our acknowledgements. It always happens in the rush to complete the manuscript and we sincerely apologize. We will acknowledge your contribution in the second, improved edition.

GUIDELINES FOR A BEGINNER

If you are a genuine beginner you might find the following guidelines useful.

Read Chapter 1 at leisure — you will come back to it again and learn the history and more about superconducting materials and applications in the course of your study.

Study Chapters 2 and 3 in detail and try to answer all the questions given at the end of the book. Do not expect to fully understand everything the first time; we recommend that you occasionally reread these chapters after studying the rest of the book. Chapter 5 gives results of the microscopic theory and opens a link with graduate textbooks. Together with Chapters 2 and 3 it provides an elementary understanding of superconductivity.

Chapters 4 and 6 can be skipped initially. Chapter 4 will give you a better understanding of the properties of the mixed state and critical current of type-II superconductors which is important for high current and high field applications of these materials. If you have an interest in metrology and electronics applications, Chapter 6 provides an elementary introduction.

Chapter 7 has to be studied slowly and reread several times. Do not expect to understand oxides after the first or even after the third reading! This is a field of active research and many questions are still open.

Finally, read Chapter 8 at leisure and reread Chapter 1! And start again ... . Remember that only by constant repetition and in-depth study of the key concepts of superconductivity at somewhat higher level can one gradually gets familiar with this fascinating subject.

Finally, if you ever get discouraged or stuck, talk to a friend, discuss with an experienced scientist who will explain things by using further examples or simply — write to the authors.

CONTENTS

Chapter 1. INTRODUCTION: Superconductivity and Superconducting Materials

Preview

In this chapter we present simple introductory notions on superconductivity, superconducting materials and their applications.

1.1. What is a Superconductor?

For a material to be considered a superconductor it has to exhibit two distinctive properties:

1)   **No resistivity:**

$$\rho = 0 \ \text{ for all } \ T < T_c \ .$$

Zero resistivity, i.e., infinite conductivity, is observed in a superconductor at all temperatures below a critical temperature, T_c (see Figure 1.1). However, if we pass a current higher than the critical current density J_c, superconductivity disappears.

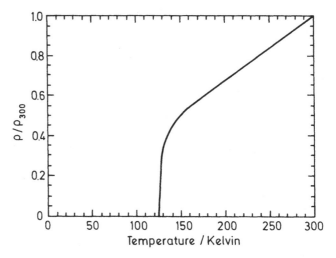

Figure 1.1: Temperature dependence of electrical resistivity of the oxide superconductor, $Tl_2Ba_2Ca_2Cu_3O_{10}$.

Is the resistivity of a superconductor really zero?

Yes. The resistivity of a superconductor to direct current is zero as far as it can be measured. A striking way to demonstrate zero resistivity is to induce a current around a closed ring of a superconducting metal. Experiments have been performed in which a 'persistent current' has run for over two and a half years without any measurable decay. This implies that the resistivity of a superconductor is smaller than 10^{-23} Ωm which is some 18 orders of magnitude smaller than the resistivity of copper at room temperature.

Why is the resistivity of a superconductor zero?

This is a non-trivial question which we shall discuss in Chapter 5. To satisfy our curiosity we shall give here a simple phenomenological description similar to the one found in the first section of Chapter 2.

When we cool a metal like aluminum below the critical temperature T_c, the gas of 'repulsive' individual electrons that characterizes the normal state transforms itself into a different type of fluid: a quantum fluid of highly correlated **pairs of electrons** (in the reciprocal *momentum space*, <u>not</u> in a real space). A conduction electron of a given momentum and spin gets weakly coupled with another electron of exactly the opposite momentum and spin. These pairs are called **Cooper pairs**. The 'glue' is provided by lattice elastic waves, called phonons.

The behavior of such a fluid of correlated Cooper pairs is different from the normal electron 'gas': they all move in a single coherent motion. A local perturbation, like an impurity, which in the normal state would scatter conduction electrons (and cause resistivity), cannot do so in the superconducting state without immediately affecting the ensemble of Cooper pairs that participate in the collective superconducting state. Once this collective, highly coordinated, state of coherent 'super-electrons' (Cooper pairs) is set into motion (like the supercurrent induced around the loop), its flow is without any dissipation. There is no scattering of 'individual' pairs of the coherent fluid, and therefore no resistivity.

2)   **No magnetic induction**:

$$B = 0 \quad \text{inside the superconductor .}$$

The magnetic inductance becomes zero inside the superconductor when it is cooled below T_c in a weak external magnetic field: the magnetic flux is expelled from the interior of the superconductor (see Figure 1.2).

This effect is called the **Meissner-Ochsenfeld effect** after its discoverers and it is the ultimate practical test in any new material. Let us also note that there always exists some critical field, H^*, above which superconductivity disappears.

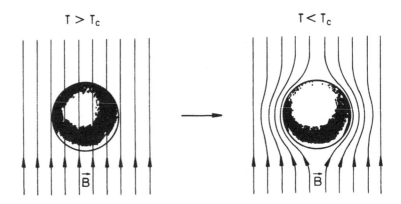

Figure 1.2: Expulsion of a weak, external magnetic field from the interior of the superconducting material.

1.2. Brief Historical Introduction

In **1911** the Dutch physicist, Heike Kamerlingh Onnes, discovered in his laboratory in Leiden that the dc resistivity of mercury suddenly drops to zero whenever the sample is cooled below 4.2 K, the boiling point of liquid helium. He named the new phenomenon – **superconductivity**. A year later, Onnes discovered that a sufficiently strong magnetic field restores the resistivity in the sample; so does sufficiently strong electric current. As we shall see, this restricts the applications of superconductivity to only a few selected materials.

In the years to follow it was discovered that many other metallic elements exhibit superconductivity at very low temperatures (see Table 1.1). Superconductivity was discovered in 1913 in lead ($T_c = 7.2$ K), while in **1930** the highest critical temperature of all pure metals was discovered in Nb, $T_c = 9.2$ K.

In **1933** Meissner and Ochsenfeld discovered another distinct property of the superconducting state: **perfect diamagnetism**. They noticed that the magnetic flux is expelled from the interior of the sample that is cooled below the critical temperature in weak external magnetic fields (see Figure 1.2).

Following the discovery of the Meissner effect, F. and H. London proposed in **1934** a simple two-fluid model. The London model (discussed in Chapter 3) explained the Meissner effect and predicted the **penetration depth** λ: this is a characteristic length of penetration of the static magnetic flux into a superconductor. While the interior of a pure superconducting metal expels the magnetic flux

Legend: Transition temperature in K / Critical magnetic field at absolute zero in gauss (10^{-4} Tesla)

1	2	3	4	5	6	7	8	9	10	11	12	13	14	15	16	17	18
Li	Be 0.026											B	C	N	O	F	He
Na	Mg											Al 1.140 / 105	Si	P	S	Cl	Ar
K	Ca	Sc	Ti 0.39 / 100	V 5.38 / 1420	Cr	Mn	Fe	Co	Ni	Cu	Zn 0.875 / 53	Ga 1.091 / 51	Ge	As	Se	Br	Kr
Rb	Sr	Y	Zr 0.546 / 47	Nb 9.50 / 1980	Mo 0.92 / 95	Tc 7.77 / 1410	Ru 0.51 / 70	Rh 0003 / 049	Pd	Ag	Cd 0.56 / 30	In 3.4035 / 293	Sn(w) 3.722 / 309	Sb	Te	I	Xe
Cs	Ba	La 6.00 / 1100	Hf 0.12	Ta 4.483 / 830	W 0.012 / 1.07	Re 1.4 / 198	Os 0.655 / 65	Ir 0.14 / 19	Pt	Au	Hg(α) 4.153 / 412	Tl 2.39 / 171	Pb 7.193 / 803	Bi	Po	At	Rn

Table 1.1: The transition temperature T_c and critical magnetic fields for superconducting elements (after Matthias and Geballe, adapted from Kittel 1986).

and is therefore flux free (perfect diamagnetism), the static flux persists within a sheath of depth λ at the surface of the sample; its magnitude decreases exponentially towards the core of the superconductor.

In **1950** Vitaly Ginzburg and Lev Landau proposed an intuitive, **phenomenological theory** (often called 'macroscopic theory') of superconductivity. Their approach turned out to be surprisingly successful and, as we shall see in this book, one can obtain useful insight into the characteristic properties of most interesting superconducting materials (including high-T_c oxides) by applying the results of the **Ginzburg-Landau theory** (discussed in Chapter 3). In **1957**, Alexei Abrikosov studied the behavior of superconductors in an external magnetic field and discovered that one can distinguish two types of material: **type-I and type-II super conductors**. While the former expel magnetic flux completely from their interior, the latter do it completely only at small fields, but partially in higher external fields. Thanks to this, i.e., the formation of the so-called mixed state, these materials can sustain superconductivity even in fields higher than 10 Tesla. Type-II superconductors are therefore the ones that are of interest for most large scale applications.

In **1957** John Bardeen, Leon Cooper and Robert Schrieffer proposed a complete **microscopic** theory of superconductivity that is usually referred to as the **BCS theory**. The basis of the theory is the interaction of a gas of conduction electrons with elastic waves of the crystal lattice. Ordinarily the electrons repel each other by the Coulomb force, but in the special case of a superconductor at sufficiently low temperatures there is a net attraction between two electrons that form the so-called Cooper pairs. Naively one can think of an electron that polarizes its environment, i.e., attracts positively charged ionic background of the lattice, which in turn attracts another electron of the opposite momentum. Below the critical temperature the attraction permits the formation of **Cooper pairs** that are pairs of electrons of opposite momenta and spins.

As a result of such attractive interaction, the 'condensed state' of highly correlated pairs of conduction electrons is formed below T_c. **All Cooper pairs move in a single coherent motion**, so a local perturbation, like an impurity, cannot scatter an individual pair. Once this collective, highly coordinated state of coherent 'super-electrons' is set in motion, its flow is without any dissipation.

The BCS theory still provides the basis for our present understanding of superconductivity in 'conventional' materials, and to some extent plays a role of the 'reference' theory in the on-going search for a correct description of superconductivity in high-T_c cuprate oxides.

In **1962** Brian Josephson postulated a fascinating quantum tunnelling effects that should occur when a supercurrent tunnels through an extremely thin layer (~ 10 Å) of an insulator. His predictions were confirmed within a year and the effects are now known as the **Josephson effects**. Superconducting technology based on these effects gradually evolved and today Josephson junction technology represents the basis of the promising superconducting electronics.

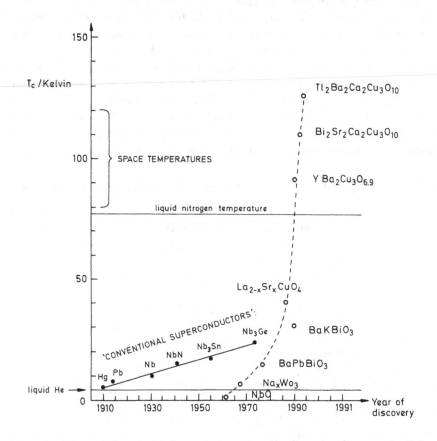

Figure 1.3: The evolution of critical temperatures since the discovery of superconductivity.

As can be seen in Figure 1.3, many technologically important Nb-compounds were discovered and thoroughly studied in the sixties and early seventies. In January **1986** Georg Bednorz and Klaus Alex Müller, researchers in IBM Laboratory at Rüschlikon, discovered superconductivity in cuprate oxides. More precisely, they found evidence for superconductivity at ~ 30 K in LaBaCuO ceramics. Following an initial disbelief of the naturally skeptic scientific community, their results were unambiguously confirmed (and even improved) in the fall of the same year. This triggered the well-publicized 'superconductivity revolution', with an outburst of research activity in thousands of laboratories all around the world (see Figure 1.3).

In February **1987** research groups in Alabama and Houston, coordinated by K. Wu and Paul Chu, discovered $Y_1Ba_2Cu_3O_7$ ceramics with $T_c = 92$ K. This was an important discovery as it meant that for the first time the world has witnessed the existence of a superconductor with a critical temperature above that of liquid nitrogen which is a much cheaper coolant than liquid helium. It is very easy to prepare $Y_1Ba_2Cu_3O_7$ ceramics by mixing, calcining and oxidizing the constituent powders so that such experiments are being done even by high school students. Only a year later, early in 1988, Bi- and Tl-cuprate oxides (see Table 1.2) were discovered with $T_c = 110$ and 125 K respectively. Since then, several research groups have claimed even higher transition temperatures, but none of them were reproducible or independently confirmed by other laboratories.

Table 1.2: The critical temperature T_c and critical magnetic field B^* for selected superconducting materials.

Conventional superconductors		
	T_c/K	B^*/T
Metallic elements:		
Al	1	0.01
Pb	7	0.08
Nb	9	0.2
Binary alloys:		
Nb-Ti	9	14
Binary compounds:		
Nb_3Sn	18	24
Nb_3Ge	23	38
Organic phases:		
κ-(BEDT-TTF)$_2$Cu(NCS)$_2$	12	20
Chevrel phases:		
$PbMo_6S_8$	15	60
High-T_c cuprate oxides		
$La_{2-x}Sr_xCuO_4$	38	40
$YBa_2Cu_3O_7$	92	>100
$Bi_2Ca_2Sr_2Cu_3O_{10}$	110	>120
$Tl_2Ca_2Ba_2Cu_3O_{10}$	125	>130

1.3. Superconducting Materials

1.3.1. Superconducting elements

As can be seen in Table 1.1 most **metallic elements** are superconductors. Their critical temperatures are typically of the order of a few Kelvins. Throughout this book we shall often refer to aluminum ($T_c = 1.1$ K) and niobium which exhibits the highest critical temperature of all the pure elements: $T_c = 9.2$ K. Note that noble metals, copper, silver and gold, and alkaline metals, sodium and potassium, all of which are excellent conductors of electricity at ambient temperatures, are not superconducting even down to very low temperatures (if at all). This very interesting fact, that superconductivity occurs in 'bad' metals (rather than in the best conducting ones), will be discussed in Chapter 5. Magnetic metals do not exhibit superconductivity.

Hydrogen, the simplest of all elements, is in a gaseous state at normal pressures. However, there are conjectures that under tremendous pressure of $\sim$ 2–3 Mbar hydrogen becomes a dense solid metallic element. Theoretical calculations based on the BCS theory predict superconductivity in dense solid hydrogen with T_c at 240 K! Helium, the next simplest of elements, is not a superconductor, but rather a superfluid below 2.2 K.

The best known semiconductors, Si and Ge, become superconductors under a pressure of $\sim$ 2 kbar with $T_c = 7$ and 5.3 K respectively. Other elements that become superconductors under pressure include P, As, Se, Y, Sb, Te, Cs, Ba, Bi, Ce, and U.

It was discovered in 1991 that a new form of solid carbon, when doped, becomes superconducting. The C_{60} molecule assumes the structure of a soccer ball with carbon atoms on a truncated icosahedron and is called buckminsterfullerene (see Figure 1.4). The C_{60} clusters crystallize to form an fcc solid. One can dope such a molecular solid with alkaline metals and observe superconductivity: $T_c = 18$ K and 30 K for K_3C_{60} and Rb_3C_{60} respectively. There are even indications of somewhat higher T_c's. By comparison, T_c's of graphite intercalated compounds with the same elements are only 0.5 K and 0.03 K for C_8K and C_8Rb respectively.

1.3.2. Binary alloys and compounds

In most alloys and compounds the critical temperatures are usually somewhat higher than in elemental metals (see Table 1.2). In this book we shall be particularly interested in Nb compounds, like Nb_3Sn, Nb_3Ge and, in particular, Nb-Ti. While the maximum current density that one can pass through the 'standard' water-cooled copper wire at 300 K is about 2000 Acm^{-2}, one can pass very high current densities

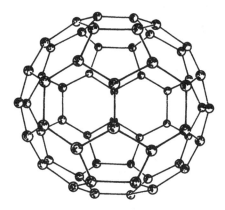

Figure 1.4: Schematic diagram of the new form of carbon, C_{60} molecule, called buckminster-fullerene (or 'buckyball'). Molecular solids of C_{60} doped with K or Cs exhibit superconductivity.

of up to 10^4 Acm^{-2} in high magnetic fields of 10 Tesla at 4.2 K through a wire made of NbTi without destroying superconductivity. This enables the construction of powerful supermagnets that provide a basis for a range of large scale applications, like energy storage or levitation of trains, for example.

Transition metals combined with other elements often produce binary alloys or compounds with T_c's higher than those of the starting elements. The intermetallic compounds and ordinary compounds usually exhibit the highest T_c's.

Despite the existence of the successful BCS theory which gave satisfactory explanation of the phenomenon of superconductivity in most metals, researchers looking for new superconducting compounds rarely relied on the 'formal' theory, which in any case never attempted to predict new materials but rather elucidated the mechanism of superconductivity. In that respect many superconductors were discovered empirically; a well known pioneer in this area was Berndt Matthias who discovered many superconducting compounds of technical importance.

Intermetallic compounds

Among the intermetallic superconductors, the most favorable group is the one based on the A_3B compound. In the cubic A-15 structure six binary compounds have T_c over 17 K. The highest T_c near 23 K is obtained in Nb_3Ge stabilized by traces of oxygen or aluminum.

The structural arrangement of the binary structure A_3B (Cr_3Si), called A-15 or β-tungsten, is given in Figure 1.5. The B atoms form a body-centered cubic sublattice; the A atoms sit on the faces of the cube to form three sets of non-intersecting orthogonal chains. The distance between the A atoms in the chains are shorter than that between chains. In these materials the A atoms are transition metals like Nb or V with unfilled d-shells, and the B atoms are mainly non-transition metals like Sn, Al, Ga, Si, Ge. The A-15 structure exists in about 70 binary compounds. Its stability towards such perturbations as pressure, temperature or stoichiometry has been studied in several systems.

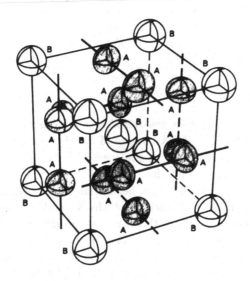

Figure 1.5: Schematic structure of an A-15, i.e. A_3B compound (after J. Müller 1980).

The presence of densely packed orthogonal chains is believed to be responsible for the crystalline instability observed in some of the A-15 superconductors (e.g. Nb_3Sn, V_3Si) at $T_m > T_c$. In these two compounds the cubic symmetry ($a = b = c$) is modified into a tetragonal one ($ab = c$) by a first order transition at $T_m = 20.5$ K in V_3Si (Figure 1.5) and at 43 K in Nb_3Sn. The change in the lattice parameter produces a change in the distance between A atoms along the [100] chain, and this lattice distortion is accompanied by a softening of acoustic phonons. Table 1.3 gives some features of superconductivity in this class of material. The high values of T_c are believed to be due to the nearby lattice instability. T_c increases when the temperature of the martensitic lattice transformation decreases.

Table 1.3: The critical temperature T_c and critical magnetic field B^* of A-15 compounds.

Compound	T_c/K	B^*/T
V_3Al	9.6	
V_3Ga	15.4	23
V_3Si	17.1	23
V_3Ge	7	
V_3Sn	4.3	
Nb_3Al	18.9	33
Nb_3Ga	20.3	34
Nb_3Si	18.0	
Nb_3Ge	23	38
Nb_3Sn	18.3	24

Chevrel phases

In 1971, Chevrel and coworkers discovered a new class of ternary molybdenum chalcogenides of the type $M_xMo_6X_8$, where M stands for a large number of metals and rare earths (RE) and X for the chalcogens: S, Se or Te. Table 1.4 gives a selection of such compounds with their values of T_c and B^*.

These compounds crystallize in a hexagonal-rhombohedral structure (Figure 1.6). It can be described by a stacking of Mo_6X_8 building units. Each of these units is a slightly deformed cube with the 8 X atoms sitting at the corners and the 6 M atoms in the centers of the faces. The M atoms form a nearly cubic lattice in which the Mo_6X_8 units are inserted. All the sites of M can be filled by large cations like Pb, Sn, rare earths (RE), or statistically, by smaller cations Cu, Ni, Fe, Co. This structure leads to materials particularly brittle, which give problems in the fabrication of wires.

Table 1.4: Critical temperatures and critical magnetic fields for several Chevrel phases.

	T_c/K	B^*/T
$SnMo_6S_8$	12	34
$PbMo_6S_8$	15	60
La Mo_6S_8	7	45
Sn Mo_6 Se$_8$	4.8	
Pb Mo_6 Se$_8$	3.6	3.8
La Mo_6 Se$_8$	11	5

The highest T_c in the series is obtained in $PbMo_6S_8$, whose value is 15 K. This compound has also an unusual high critical field of 60 T, for a conventional superconductor. The large values of the upper critical field as compared with Nb_3Sn and NbTi make these materials interesting for making wires. Critical currents as high as $\sim 3 \times 10^5$ Acm^{-2} have been observed at 4.2 K and this provides an impetus for making wires out of these very brittle materials.

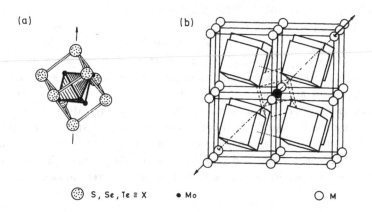

(a)   (b)

S , Se , Te ≡ X   • Mo   ◯ M

Figure 1.6: Schematic structure of a Chevrel phase, $M_x Mo_6 X_8$: a) one unit of $Mo_6 X_8$, b) stacking of 8 $Mo_6 X_8$ units into the rhombohedral unit cell. Note that the atom M is centered at the origin (after Yvon 1982).

The existence of $(RE)Mo_6 X_8$ has led to the problem of coexistence of magnetism of rare earth and superconductivity. It has been found for the first time that antiferromagnetism of the rare earth can coexist with superconductivity like in Gd, Tb, Dy, Er compounds where T_c is 1.4, 1.65, 2.1 and 1.85 K, T_N is 0.84, 0.9, 0.4 and 0.15 K respectively. In $HoMo_6 S_8$, long range magnetic order produces ferromagnetism which destroys superconductivity. This is called reentrant superconductivity. The material is superconducting only between two critical temperatures 2 K and 0.65 K. Below 0.65 K the material is ferromagnetic.

In this class of materials coexistence of the magnetic rare earth and superconductivity is possible because there is a very small interaction between the rare earth and the itinerant electrons which gives rise to superconductivity and which are mainly on the $Mo_6 S_8$ units.

1.3.3. Organic superconductors

Organic superconductors are a novel group of materials. The first organic superconductor $[TMTSF]_2 PF_6$, where TMTSF denotes tetramethyltetraselenafulvalene, was discovered by K. Bechgaard and D. Jérome in 1980 and had a T_c of 1 K. Subsequent developments have led to higher T_c materials exhibiting a variety of novel electronic and superconducting properties. These systems were characterized by their nearly one-dimensional properties and by low carrier concentration. Later, a new series of materials with a two-dimensional character was discovered: the [BEDT-TTF]$_2$X, where BEDT-TTF denotes bis-ethylenedithio-tetrathiafulvalene. The κ modification of the X = Cu(NCS)$_2$ compounds has the highest T_c of $\sim$ 10 K.

In the first series, [TMTSF]$_2$X, the planar TMTSF molecules (Figure 1.7a) form stacks along which the electrons are most conducting. Two molecules donate one electron to an anion X and an organic salt is constructed. Most of these organic compounds are superconducting only under pressure, and Table 1.5 gives critical temperatures and critical pressures at which superconductivity is observed. The most important property of these compounds is their quasi one-dimensional nature. All properties are very much anisotropic. There is a factor of 10 difference between conductivities along and perpendicular to the chains. Consequently the superconducting properties and the critical field are also very anisotropic.

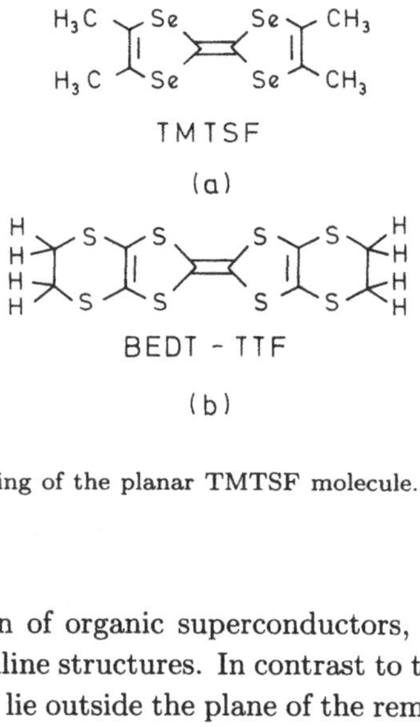

TMTSF

(a)

BEDT – TTF

(b)

Figure 1.7: a) Schematic drawing of the planar TMTSF molecule. b) Schematic drawing of the BEDT-TTF molecule.

The second generation of organic superconductors, the [BEDT-TTF] family, has a rich variety of crystalline structures. In contrast to the flatness of the TMTSF molecules, the CH$_2$ groups lie outside the plane of the remaining part of the BEDT-TTF molecule (Figure 1.7b). Among the family the β-(BEDT-TTF)$_2$X, where X is an anion such as I$_3$, IBr$_2$, AuI$_2$, has attracted the most interest (see Table 1.6). They have the highest T_c and the crystals have a layered structure which consists of an

Table 1.5: Quasi-one-dimensional organic superconductors, $(TMTSF)_2X$.

X	P_c/kbar	T_c/K
ClO_4	0	1.2
PF_6	9	1.2
ReO_4	9.5	1.4

array of BEDT-TTF stacks forming conducting layers separated by insulating anion sheets. Consequently the electronic structure is of two-dimensional nature which appears in the anisotropy of the conductivity and the superconducting properties.

Organic superconductors attract attention from the viewpoint of similarity to the high-T_c oxide superconductors. They have low dimensionality, low electron or hole concentrations, and one can study the competition between superconductivity and magnetism (Figure 1.8). Thus these materials might provide the evidence for a mechanism of superconductivity different from the conventional one.

Table 1.6: Quasi-bi-dimensional organic superconductors, $(BEDT\text{-}TTF)_2X$.

X		T_c/K
I_3	β_L	1.2
I_3	β_H	8.1
IBr_2	β	2.5
AuI_2	β	4.2
$Cu(NCS)_2$	κ	10

1.3.4. High-T_c oxides

This book was written with the aim of helping non-specialists enter the field of high-T_c oxides and their properties and applications. We dedicate the whole of Chapter 7 to these materials so at present we only emphasize some simple but important facts on these materials:

1. While ceramics are usually insulators (for example, your coffee cup), these ceramics are superconductors!

2. Their critical temperatures are ~ 100 K, i.e. an order of magnitude higher than that of conventional metals, alloys and compounds.

3. One needs very high magnetic fields to completely destroy superconductivity in these materials ($\sim$ 100 Tesla at 4.2 K). Many other fascinating properties will be discussed in Chapter 7.

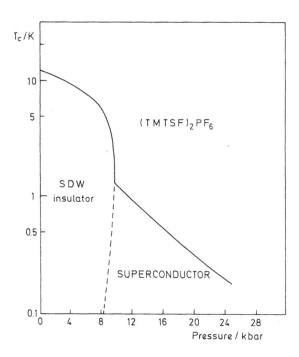

Figure 1.8: Schematic electronic phase diagram of (TMTSF)$_2$PF$_6$ (after Jerome 1990).

1.4. Applications

Already in the years following the discovery of superconductivity, Kamerlingh Onnes tried to explore the potential of zero resistivity state for electrical applications. However, when he tried to build electromagnetic coils made with wires of pure superconducting metals, like lead or indium, he was unable to do so. These metallic superconductors easily lose their superconducting properties in weak external magnetic fields (magnetic induction $B < 0.1$ Tesla). Furthermore, they were able to carry only weak electric currents. So, Kamerlingh Onnes rapidly learned that sufficiently strong external magnetic fields or strong currents destroy the superconducting state in most pure metals and bring them back to the normal state.

As we have mentioned in the previous section and we shall often repeat through-out this book, superconductors of interest to strong current applications, which can sustain very high magnetic fields, are usually fairly complex materials, like Nb compounds, Chevrel phases, or cuprate oxides.

Generally speaking, one can distinguish two main groups of applications of superconducting materials, discussed in Sections 1.4.1 and 1.4.2.

1.4.1. Large scale applications

From the practical point of view the most useful aspect of conventional type-II superconductors (predominantly Nb compounds) is their capacity to carry high transport currents with acceptably low energy dissipation. For example, one could construct a 10-Tesla superconducting magnet using copper wire, but such a magnet would dissipate 2 MW of power. Therefore one mostly uses Nb-compounds to construct commercial magnets. Wires made with these materials remain superconducting at 4.2 K even in magnetic fields considerably higher than the required performance peak of the magnet. Furthermore, for equal power and the uppermost field, the superconducting magnet is much smaller than its normal counterpart made out of copper wires.

Powerful superconducting magnets, **supermagnets**, are used in medicine for diagnostics, for example in nuclear spin tomographs, for energy storage, for magnetic levitation (trains), and in research laboratories, particularly in high-energy particle accelerators for fundamental physics research.

1.4.2. Superconducting electronics applications

These are mainly thin film applications in ultra-fast microelectronics or instrumentation.

Thin film applications in electronics, together with superconducting magnets, are usually considered the most important area of superconductivity-based technology. Most of these applications are based on **Josephson effect**, a quantum phenomenon that enables the construction of the fastest nanoscopic switches, Josephson junctions, and related device structures, **SQUIDs**: Superconducting Quantum Interference Devices.

Josephson junctions are sophisticated sandwich structures of superconducting films (usually of Nb) separated by extremely thin (~ 10 Å) insulating oxide layers. In a suitable submicroscopic circuit they act as the fastest switching elements available today.

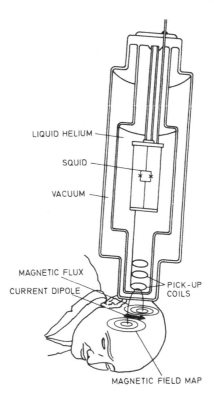

LIQUID HELIUM

SQUID

VACUUM

MAGNETIC FLUX

CURRENT DIPOLE

PICK-UP COILS

MAGNETIC FIELD MAP

Figure 1.9: Schematic diagram of a SQUID-based unit used in medical diagnostics (courtesy of Scientific American 1989).

Josephson junctions have also the lowest power consumption at their operating temperature, so they might provide the basis for the architecture of some of the fastest computers of the future. They are already used in the fastest commercially available oscilloscopes that operate at 10 GHz.

As the name suggests, SQUIDs explore subtle quantum interference effects: an analysis of superconducting loop shows that the magnetic flux that can thread the loop is quantized in quantum units of flux which have a value of 2×10^{-15} Weber. SQUIDs are suitably processed superconducting loops which detect minute changes in magnetic flux, i.e. they are high-sensitivity magnetic flux detectors that can be used in the finest precision instruments at the forefront of metrology.

Summary

1. For a material to be considered as a superconductor it has to exhibit two distinctive hallmarks of superconductivity:

i) $\rho = 0$ for all $T < T_c$.

Zero resistivity, i.e., infinite conductivity is measured in a superconductor below some critical temperature, T_c.

ii) $\mathbf{B} = 0$ inside the superconductor .

The **Meissner effect**: in a weak external magnetic field below T_c magnetic flux is expelled from the interior of the superconductor.

2. Pure metals exhibit T_c's of a few Kelvins, up to ~ 9.2 K in Nb, while T_c's of binary metallic alloys are somewhat higher with a record of 23 K in Nb$_3$Ge. The most relevant for high current applications are Nb-Ti ($T_c = 9$ K), Nb$_3$Sn ($T_c = 18$ K) and Chevrel phases like PbMo$_6$S$_8$ ($T_c = 15$ K). These materials are usually referred to as conventional superconductors.

Critical temperatures above liquid nitrogen temperature ($T_c > 77$ K) were discovered in several high-T_c oxides with $T_c \sim 100$ K.

3. Some of the most important applications of superconductors are either in high current **wires** for powerful **supermagnets** or in thin film structures like **Josephson junctions** for superconducting electronics. SQUIDs have been used to develop some of the most sensitive measurement instruments at the forefront of present technology.

Further Reading

Jonathan L. Mayo: *Superconductivity — The Threshold of a New Technology*, TAB Books Inc., Blue Ridge Summit, PA-USA, 1989

Randy Simon and Andrew Smith: *Superconductors–Conquering Technology's Frontier*, Plenum Press, New York, 1988

A. M. Wolsky, R. F. Giese and E. J. Daniels: *The New Superconductors: Prospects for Applications*, Scientific American, Feb. 1989, p. 45

Superconductors, TIME, May 11, 1987, p. 38

Chapter 2. CHARACTERISTIC PROPERTIES

Preview

In this chapter we discuss the origin of electrical resistivity in the normal metal and contrast it with the absence of resistivity in the superconductor. We introduce the basic concept of the **superconducting wavefunction** which will be used throughout the whole book. We then show that the Meissner effect is another distinct characteristic property of the superconducting state. We emphasize the difference in behaviors in external magnetic fields between type-I and type-II superconductors, and then we briefly discuss critical currents in type-II materials. We give a short overview of other characteristic features of the superconducting state. The existence of the energy gap implies an **energy scale** which permits to understand the response of a superconductor to high frequency electromagnetic fields. We end the chapter with flux quantization and Josephson effect that clearly illustrate the consequences of the description of superconductivity by superconducting wavefunctions.

2.1. Normal Metal vs. Superconductor

2.1.1. Description of the normal state

We are all familiar with a copper wire. To understand what happens in a normal metal we shall use a simple model. It consists of a regular crystalline lattice of positively charged ions and a gas of free, non-interacting conduction electrons that fill the inter-ionic space of the lattice. If there is typically one electron per ion, we should have $\sim 10^{23}$ electrons/cm^3. As the electrons are of the opposite charge from the ions, the total charge is balanced and at equilibrium our model-metal is electrically neutral.

If we apply an electric field as an external perturbation to the gas of free electrons within the metal, the external force will accelerate the electrons and create a current flow of 'free' electrons. As the ions are arranged in perfectly regular array, they do not scatter conduction electrons at $T = 0$.[†] The scattering of electrons at $T = 0$ is actually caused by deviations from the ideal periodic potential of the lattice, i.e., by impurities, imperfections in periodicity like dislocations etc.... .

As every real metal contains some imperfections and impurities, one observes some finite resistivity at very low temperatures. This resistivity, extrapolated to $T = 0$, is called residual resistivity, ρ_i. As we increase the temperature, the electrons also get scattered by thermal vibrations of the lattice (called phonons) so the resistivity rises with temperature. This contribution is called phonon resistivity,

---

[†]If the crystal were perfect, at $T = 0$, the electron waves would propagate without scattering and there would be no resistivity, i.e., conductivity of an ideal crystal at $T = 0$ should be infinite.

$\rho_P(T)$. Therefore the temperature dependence of resistivity of a good metal, like silver, at low temperatures can be described as

$$\rho(T) = \rho_i + \rho_P(T) \; . \tag{2.1}$$

This is an empirical rule which provides a basis for understanding the resistivity of metals and alloys at low temperatures (Figure 2.1).

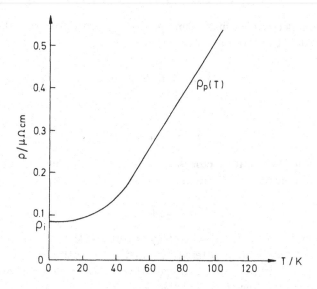

Figure 2.1: Temperature-dependent resistivity of Ag with 0.02 at %Sn at low temperatures (after Dugdale 1976).

In order to derive a simple expression for the residual resistivity of the metal we shall first familiarize ourselves with some characteristic quantities of the normal state. At $T = 0$ the maximum kinetic energy of an electron inside the metal is called Fermi energy, E_F. It is related to the number of carriers per unit volume, n, by the simple relation:

$$E_F = \frac{\hbar}{2m}(2\pi^2 n)^{\frac{2}{3}} \; , \tag{2.2}$$

where $\hbar$ is the Planck constant, h, divided by 2π and m is the mass of the electron. The Fermi energy of a typical metal is of the order of the electron volt (see Table 2.1).

Table 2.1: Some characteristic quantities of the normal state.

Material	n	v_F	l_e	$\rho(100 \text{ K})$
	10^{21} cm^{-3}	10^6 ms^{-1}	nm	$\mu\Omega$cm
Al	180	2.0	130	0.3
Nb	56	1.4	29	3
$La_{2-x}Sr_xCuO_4$	5	0.1	~ 5	~ 100
$YBa_2Cu_3O_7$	7	0.1	~ 10	~ 60

Conduction electrons of maximum energy, E_F, propagate with the Fermi velocity v_F related to the Fermi momentum $\boldsymbol{p}_F$ by

$$\boldsymbol{p}_F = m\boldsymbol{v}_F \ . \tag{2.3}$$

We have

$$E_F = \frac{1}{2}p_F v_F \ . \tag{2.4}$$

We also define the Fermi wave vector, k_F; as in quantum mechanics a wave is always associated with a particle by de Broglie relation

$$\boldsymbol{p}_F = \hbar\mathbf{k}_F \ . \tag{2.5}$$

The conduction electrons that propagate through the crystal with a characteristic Fermi velocity v_F are scattered by impurities or lattice imperfections. This gives rise to a resistivity. Between two scattering events an electron covers on average a characteristic distance l_e, called the electron mean free path. The resistivity ρ_i of a metal, according to elementary metal theory, is given by

$$\rho_i = \frac{mv_F}{ne^2 l_e} \ , \tag{2.6}$$

where e and m represent the charge and mass of the electron. In isotropic metals, the conductivity is equal to the inverse of the resistivity; both quantities are tensors in the anisotropic case.

Equation (2.6) shows that in the normal state of a given metal the resistivity is inversely proportional to the electron mean free path. The shorter the average distance between the scattering events the higher is the resistivity. The introduction of impurities into a metal obviously reduces l_e and increases ρ_i. This can be clearly seen in Table 2.1, in which typical values for ρ and l_e for several superconducting materials are presented.

2.1.2. The superconducting state:
No 'individual' scattering — no resistivity

As we have seen in Chapter 1, the electrical resistivity in superconductors is zero for temperatures below the critical temperature. So, one can apply dc electrical current (supercurrent) without loss. Let us see what happens in a superconducting state and what are its characteristic properties compared with the normal state.

Consider aluminum. In the normal state above the critical temperature ($T_c = 1.1$ K) Al is a good conductor and behaves just like our aforementioned metal-model or like copper which exhibits no superconducting behavior down to the lowest attainable temperature. Its conduction electrons behave like a gas of nearly free electrons that are scattered by lattice vibrations, lattice imperfections, etc. ..., all of which contribute to the resistivity.

However, when we cool aluminum to below T_c, its dc resistance abruptly vanishes. The resistivity is zero. One naturally wonders what happened to the scattering of conduction electrons which contributed to the resistivity in the normal state? Why did it disappear? A satisfactory explanation to these puzzling questions can be given only within the rather involved quantum mechanical description of the microscopic BCS theory, which we shall briefly discuss in Chapter 5. For time being we give only some intuitive, and rather crude, description of this phenomenon.

When we cool a metal like Al below the critical temperature T_c, the gas of the 'repulsive' individual electrons that characterizes the normal state transforms itself into a different type of fluid: a quantum fluid of highly correlated **pairs of electrons** (in the reciprocal, *momentum space*, <u>not</u> in a real space). Below T_c a conduction electron of a given momentum and spin gets weakly coupled with another electron of exactly the opposite momentum and spin. These pairs are called **Cooper pairs**. The 'glue' is provided by the elastic waves of the lattice, called phonons. One can visualize this attraction by a naive real-space picture: as the lattice consists of positive ions, the moving electron creates a lattice distortion. Due to the heavy mass of lattice ions, this positively charged distortion relaxes slowly and is therefore able to attract another electron. The 'distance' between the two electrons of the Cooper pair, called the **coherence length**, ξ , is large in metals. It has a value $\xi = 16000$ Å in pure Al, $\xi = 380$ Å in pure Nb, for example. So while the 'partners' in the Cooper pair are far apart, the other nearest electrons (belonging to other Cooper pairs of the collective state) are only a few angstroms away. The behavior of such a fluid of correlated Cooper pairs is different from the normal electron 'gas'.

The electrons which form the pair have opposite momenta (and opposite spins), so the net momentum of the pair is zero. By the de Broglie relation, Eq.(2.5), the associated wave has an infinite wavelength (physically, the wavelength is actually of the order of the size of the sample). This shows that superconductivity is a quantum phenomenon on the macroscopic scale. In optics a wave is diffused only if there is fluctuation of the number of scattering centers within a volume of the size of the wavelength. Clearly this is not the case, hence the Cooper pairs cannot be scattered by the usual scatterers of individual electrons, i.e., there is no mechanism which could give rise to resistivity.

2.1.3. The wavefunction of the superconducting state

The Cooper pair has twice the charge of a free electron, $q = 2e$. The electrons are fermions and obey the Fermi-Dirac statistics and Pauli exclusion principle which allows only one electron in a given quantum state. Cooper pairs are quasi-bosons, obey the Bose-Einstein, statistics and are allowed to be (all) in the same state. In contrast to the normal metal in which each electron has its own wavefunction, in a superconductor, all Cooper pairs are described by the single wavefunction

$$\Psi(\mathbf{r}) = \sqrt{n_s}(\mathbf{r})e^{i\varphi(\mathbf{r})} \ , \tag{2.7}$$

where $n_s(\mathbf{r})$ can be considered as the number of 'superconducting electrons' (Cooper pairs). Note that $\Psi(\mathbf{r})\Psi^*(\mathbf{r}) = n_s(\mathbf{r})$ and $\varphi(\mathbf{r})$ is a spatially varying phase. Why did we introduce the phase? In optics, when we have a beam of photons all in the same state, i.e., traveling with the same velocity, it is described by a plane wave, $e^{ikr-i\omega t}$, and the gradient of the phase is related to the momentum of the particle by the de Broglie relation, $\mathbf{p} = \hbar\mathbf{k}$, or $\mathbf{v} = \frac{\hbar}{m}\nabla\varphi$. As all Cooper pairs are in the same state, we have an analogous situation and the gradient of the phase becomes a macroscopic quantity, a quantity proportional to the current flowing in the superconductor.

2.2. The Meissner-Ochsenfeld Effect

In addition to zero resistivity (i.e., infinite conductivity), the superconductor exhibits another striking property: it *expels the magnetic field from its interior.* Note that this is not a consequence of infinite conductivity, but of another intrinsic characteristic property of the superconducting state which we shall now discuss in some detail.

As was already illustrated in Figure 1.2, in the normal state, at temperatures above T_c the field lines pass through the metallic specimen. Upon cooling below T_c, a phase transition into the superconducting state takes place and the magnetic flux gets expelled out of the interior of the metallic sample.

The Meissner-Ochsenfeld effect cannot be deduced from the infinite conductivity of a superconductor. The exclusion of the magnetic field from the interior of a superconducting specimen is direct evidence that the superconducting state is not simply one of zero resistance. If it were so, then a superconductor cooled in the magnetic field through T_c would have trapped the field in its interior. When the external field is removed, the persistent induced eddy currents would nevertheless preserve the trapped field in the interior of the specimen. The expulsion of the flux therefore implies that this new superconducting state is a true thermodynamic equilibrium state.

The above argument can be strengthened by a few elementary formulae of electrodynamics.

Consider Ohm's law, $V = RI$, written as

$$\mathbf{E} = \rho \mathbf{J} , \tag{2.8}$$

where $\mathbf{E}$ represents the electric field, ρ the resistivity and $\mathbf{J}$ the electrical current density in the sample. Zero resistivity implies zero electric field.

So, if we take the Maxwell equation

$$\text{curl } \mathbf{E} = -\frac{\partial \mathbf{B}}{\partial t} , \tag{2.9}$$

we have

$$\frac{\partial \mathbf{B}}{\partial t} = 0 . \tag{2.10}$$

We see that the magnetic induction in the interior of the sample has to be constant as a function of time. The final state of the sample would have been different if it were cooled under an applied external field or if the field were applied after the sample has been cooled below T_c. In the former case the field would have remained within the sample, while in the latter it would have been zero. For the specimen to be in the same thermodynamic state, independent of the precise sequence that one uses in cooling or in applying the field, the superconducting metal always expels the field from its interior, and has $\mathbf{B} = 0$ in its interior. So the expulsion of the magnetic field ensures that the superconducting state is a true thermodynamic state.

2.3. Destruction of Superconductivity by Magnetic Field

Below T_c the superconducting state has lower free energy than the normal state but it requires the expulsion of the flux. This in turn costs some magnetic energy which has to be smaller than the condensation energy gained in undergoing the phase transition into the superconducting state (i.e., by forming the coherent ensemble of Cooper pairs from the 'random' electron gas). Obviously, if we begin to increase the external magnetic field it will reach the point where the cost in magnetic energy will outweigh the gain in condensation energy and the superconductor will become partially (in a particular sample geometry) or totally normal.

Superconductivity disappears and the material returns to the normal state if one applies an external magnetic field of strength greater than some critical value B_c, called the critical thermodynamic field.

The superconducting state can also be destroyed by passing an excessive current through the material, which creates a magnetic field at the surface of strength equal to or greater than B_c. This limits the maximum current that the material can sustain and is an important problem for applications of superconducting materials.

2.3.1. Type-I superconductors

Superconducting materials that completely expel magnetic flux until they become completely normal are called **type-I superconductors**. In the older literature they were often referred to as 'soft' or 'pure' superconductors. With the exception of V and Nb, all superconducting elements and most of their alloys in the 'dilute limit', are type-I superconductors. The strength of the applied magnetic field required to completely destroy the state of perfect diamagnetism in the interior of the super-conducting specimen is called the thermodynamic critical field B_c. As schematically shown in Figure 2.2, the variation of the critical field B_c with temperature for type-I superconductor is approximately parabolic:

$$B_c = B_0 \left\{ 1 - \left(\frac{T}{T_c} \right)^2 \right\} , \tag{2.11}$$

where B_0 is the extrapolated value of B_c at $T = 0$.

The magnetization curve of an ideal superconductor is given in Figure 2.3a. In the MKSA system of units one can write

$$\mathbf{B} = \mu_0(\mathbf{H} + \mathbf{M}) \tag{2.12}$$

where $\mathbf{M}$ is the magnetization and $\mu_0 = 4\pi \times 10^{-7}$.

The Meissner effect, $\mathbf{B} = 0$, corresponds to $\mathbf{M} = -\mathbf{H}$. Above the critical field B_c the material becomes normal so $\mathbf{M} = 0$. The negative sign shows that the sample becomes a perfect diamagnet that excludes the flux from its interior by means of surface currents.

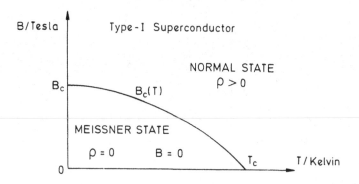

Figure 2.2: Schematic diagram of $B_c(T)$ for type-I superconductor.

2.3.2. Type-II superconductors

For a type-II superconductor there are two critical fields: the lower B_{c1} and the upper B_{c2}. The flux is completely expelled only up to the field B_{c1}. So in applied fields smaller than B_{c1}, the type-II superconductor behaves just like a type-I superconductor below B_c. Above B_{c1} the flux partially penetrates into the material until the upper critical field, B_{c2}, is reached. Above B_{c2} the material returns to the normal state (see Figure 2.3b).

Between B_{c1} and B_{c2} the superconductor is said to be in the **mixed state**. The Meissner effect is only partial. For all applied fields $B_{c1} < B < B_{c2}$, magnetic flux partially penetrates the superconducting specimen in the form of tiny microscopic filaments called vortices (see Figure 2.4).

The diameter of a vortex in conventional superconductors is typically 100 nm. It consists of a normal core, in which the magnetic field is large, surrounded by a superconducting region in which flows a persistent supercurrent which maintains the field within the core.

Each vortex carries a magnetic flux

$$\Phi_0 = \frac{h}{2e} = 2.067 \times 10^{-15} \text{ Weber },\qquad (2.13)$$

where h is the Planck constant and e is the charge of the electron. The magnetic induction B is directly related to n, the number of vortices per m^2:

$$B = n\Phi_0 . \qquad (2.14)$$

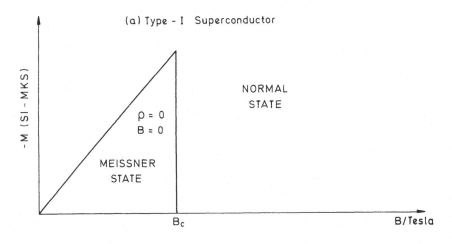

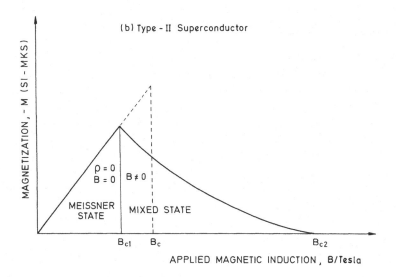

Figure 2.3: Variation of magnetization as a function of the magnetic field for (a) type-I superconductor and (b) type-II superconductor.

Thanks to this partial flux penetration the material can withstand strong applied magnetic fields without returning to the normal state. Superconductivity can and does persist in the mixed state up to the upper critical field, B_{c2}, which is sometimes as high as 60 Tesla (Chevrel phases) or even $\sim$ 150 Tesla in high-T_c oxides. At fields higher than B_{c2} the superconductor returns to the normal state.

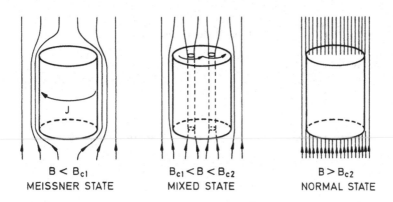

Figure 2.4: Flux penetration in the mixed state (after Decroux and Fischer 1989).

2.4. Critical Currents in Type-II Superconductors

Naïvely one would think that the critical currents create the field B_{c2} at the surface of the sample. Actually the critical currents correspond rather to those that generate the field B_{c1}, and not B_{c2}. Why does B_{c1} play a role? Because above B_{c1}, there are vortices in the material. The current displaces the vortices and this creates a non-desirable energy dissipation. The vortex in motion creates an electric field

$$\mathbf{E} = \frac{d\mathbf{\Phi}}{dt} \; . \tag{2.15}$$

In the presence of this field, the current $\mathbf{J}$ dissipates energy $\mathbf{E} \cdot \mathbf{J}$. This energy dissipation is equivalent to resistivity. Theoretically critical currents of type-II superconductors are weak; still weaker for small values of B_{c1} compared with B_{c2} which is rather large. So, how does one pass intense currents without dissipation above B_{c1}? The answer is that one has to prevent the motion of vortices so that the critical current would not be limited by B_{c1}. This is achieved by the so-called vortex pinning (or flux pinning).

How does one pin the vortex? Simply by creating sites out of which the vortex cannot leave without large energy increase. For example, in conventional type-II materials one can find small inclusions of normal metal imbedded in the superconductor. The vortex will be pinned to such inclusion as it does not have to spend energy to destroy superconductivity in that inclusion. What is the typical size of efficient inclusions? Evidently the coherence length ξ, the diameter of the tube which is the normal state within the vortex. Inhomogeneities over distances of the order of the coherence length are therefore responsible for the attainment of very high currents in some materials like Nb-Ti. In high-T_c oxide superconductors where

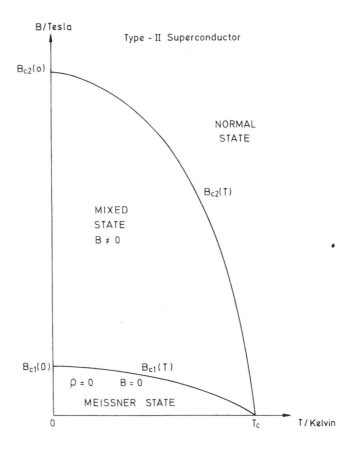

Figure 2.5: Variation of critical fields, B_{c1} and B_{c2}, as a function of temperature. The upper critical field B_{c2} can be very high, even above 100 Tesla (see Table 2.2).

the coherence length is very short ($\xi \sim 10$ Å), it is not obvious how to control the vortex pinning sites.

Practically all technologically interesting materials like Nb compounds, Chevrel phases or high-T_c oxides are type-II superconductors. Crudely speaking, the reason for this is that the creation of vortices keeps the magnetic energy smaller than the condensation energy, so the overall free energy of the mixed superconducting state remains (thermodynamically) more favorable than the normal state even up to high magnetic fields. Since the supercurrent can flow in the mixed state through the superconducting regions between vortices, type-II superconductors allow one to construct wires needed for high field magnets.

Contrary to type-I superconductors where the critical currents are rather low, the critical current density J_c can be sufficiently high in many type-II materials to enable practical applications. For example, NbTi compounds have $J_c \sim 10^6$ Acm^{-2} at 4.2 K and are used for making wires. If the upper critical field is very high (~ 14 T for NbTi, ~ 60 T for Chevrel phase), such wires can be used for construction of superconducting magnets that can and do produce magnetic fields of tens of Teslas. The physics of the mixed state, technological materials and applications, will be discussed in some detail in Chapters 4 and 8 respectively.

2.5. Thermal Properties of Superconductors

2.5.1. Heat capacity

The entropy of the superconducting metal decreases considerably upon cooling below T_c. As entropy measures the degree of disorder of a given system, this decrease signifies that the superconducting state is more ordered than the normal state. The fraction of electrons that is thermally excited in the normal state becomes ordered in the superconducting state. The entropy variation is relatively weak, of the order of 10^{-4} k_B per atom (k_B is the Boltzmann constant).

The transition to the superconducting state is accompanied by quite drastic changes in the thermodynamic equilibrium and thermal transport properties of the superconductor. In particular, the heat capacity of the superconductor changes at T_c in a characteristic way (see Figure 2.6). In zero magnetic field there appears a

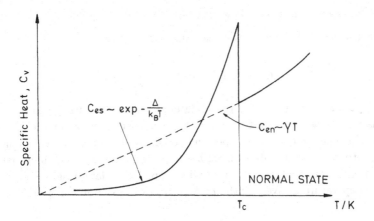

Figure 2.6: Schematic presentation of the heat capacity of a metal in the normal and the superconducting state. Note the characteristic jump at the transition point.

discontinuity at T_c. At temperatures immediately below T_c, the heat capacity is much larger than in the normal state so the sudden 'jump', together with the more rapid decrease with decreasing temperature, give rise to this characteristic shape below the transition.

2.5.2. Thermal conductivity

In normal metals like copper, large electrical conductivity is accompanied by large thermal conductivity; the ratio between the two is approximately constant (Wiedemann-Franz law). However, in the superconducting state the thermal conductivity is smaller than in the normal state and almost vanishes at very low temperatures. This interesting property can be used in low temperature technology for 'heat switches': thermal contact between two materials established by a superconducting material can be controlled by a magnetic field which switches off the superconductivity when heat transfer is needed, or alternatively preserves it (by turning off the magnetic field) when the thermal separation is preferable.

Intuitively one can understand this unusual combination of infinite electrical conductivity and small thermal conductivity as follows: the transport of heat requires the transport of entropy (which measures disorder). The superconducting state is one of perfect order (zero entropy) so there is 'no disorder to transport'; hence the vanishing thermal conductivity due to electrons. On the contrary, small thermal conductivity due to phonons can even increase in the superconducting state.

2.5.3. The energy gap and the characteristic energy scale

At temperatures well below T_c the heat capacity varies as

$$C_v(T < T_c) \sim \exp\left(-\frac{\Delta}{k_B T}\right).$$ (2.16)

$E_g = 2\Delta$ is a constant for a given material, called **the energy gap**; k_B is the Boltzmann constant. Such a temperature dependence is characteristic of a system that has an energy gap in its spectrum of allowed energy states (Figure 2.7). Although there are materials with gapless superconductivity, most materials of interest to us in this book do have an energy gap. The relation between Δ and the critical temperature is given by the BCS theory (discussed in Chapter 5):

$$2\Delta = 3.5 k_B T_c.$$ (2.17)

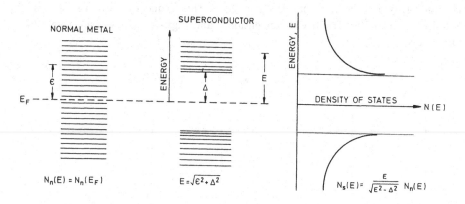

Figure 2.7: Schematic density of states of a normal metal and of a superconductor (after Bardeen 1990).

By using a simple temperature-energy conversion formula, 1 eV $\sim$ 12000 K, it is easy to estimate that the gap in conventional superconducting materials with $T_c < 20$ K is of the order of 1 meV, while in high-T_c oxides with T_c of $\sim$100 K, $\Delta \sim$ several (1–10) meV. To those familiar with elementary semiconductor physics, these superconductor energy gaps will seem rather small as compared with $E_g \sim 1.5$ eV in GaAs ($\sim$1.2 eV in Si at $T = 0$).

However, while in semiconductors the gap in the energy spectrum corresponds to the energy difference between the valence and the conduction band and is therefore on the scale of ~ 1 eV, in the superconductor 2Δ corresponds to the energy needed to break a Cooper pair. While in semiconductors, like GaAs, the electron-hole recombination across the gap releases photons (quanta of light radiation) of nearly 1 μm in wavelength, weakly coupled Cooper pairs can be broken by less energetic individual photons of longer wavelengths: 0.1 to 1 mm (1 eV = 1.24 μm).

2.6. High-frequency Electromagnetic Properties

For all frequencies much higher than the frequency corresponding to the energy gap

$$E_g = h\nu \, , \tag{2.18}$$

where ν is the frequency in Hz, the electromagnetic response in the superconducting state is identical to the response of the normal state. If we remember that

$$1 \text{ eV} \sim 10^{14} \text{ Hz} \, , \tag{2.19}$$

we can easily understand that the change in the frequency response occurs at $\nu \sim 10^{11}$ and $\sim 10^{12}$ Hz in the conventional and high-T_c oxides respectively (see Figure 2.8).

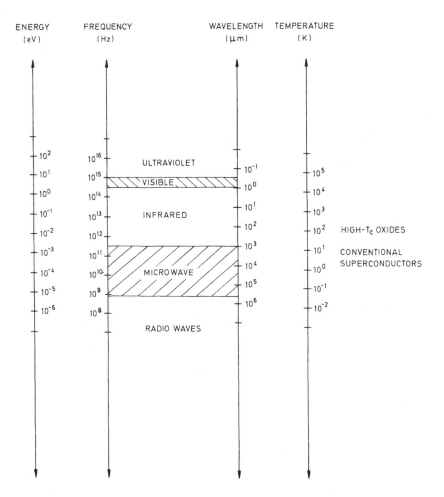

Figure 2.8: A chart of the electromagnetic spectrum and characteristic energies of conventional and high-T_c superconductors.

High-frequency electromagnetic properties of superconductors differ from zero frequency (or very low frequency) behavior discussed so far. In the radio-frequency ($< 10^8$ Hz) and microwave frequency range ($\sim 10^8$–10^{11} Hz) the resistance of the superconductor to current flow is not zero. However, the resistance and the accompanying energy loss are still rather small.

In the optical region of the spectrum ($\sim 10^{15}$ Hz) there is no difference in electromagnetic response between the normal and the superconducting state so one does not see any change in the appearance of the sample as it undergoes the transition. In the range of $\sim 10^{11}$ to 10^{12} Hz (depending on the material) there is a sharp increase in the absorption of electromagnetic radiation by the superconductor. This is due to the existence of the aforementioned energy gap in the electronic energy spectrum.

The sharp rise in the absorption occurs at the energy (Planck constant times the frequency, $E = h\nu$) of a single photon which is just sufficient to produce an excitation (by depairing weakly coupled Cooper pairs).

2.7. Characteristic Phenomenological Parameters

2.7.1. The magnetic penetration depth

The Meissner effect shows that $B = 0$ in the interior of a superconductor. This, however, cannot be true at the surface of the superconductor. Actually, to cancel B one requires currents on the surface which give rise to magnetization M so that in the interior of the superconductor $M + H = 0$. As the resistivity is zero these surface currents do not dissipate energy. Therefore we call them superfluid currents or simply **supercurrents**. The thickness of the region of the sample (measured from the surface) through which flow the supercurrents is called the **penetration depth** of the magnetic field, λ. It is one of the characteristic quantities (lengths) that characterize the superconductor.

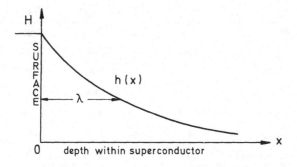

Figure 2.9: Schematic decay of the magnetic field in the interior of a superconductor.

The penetration length is also the extension of the supercurrent around the aforementioned flux tubes (vortices) that penetrate into the sample in the mixed state (between B_{c1} and B_{c2}) of type-II superconductors.

2.7.2. The coherence length

In Section 2.1 we referred to the coherence length ξ as the distance between two electrons of the Cooper pair within the highly correlated coherent superconducting state. This 'definition' was only an introductory one and throughout this book we shall re-define the coherence length more and more rigorously.

Another introductory 'definition' of the coherence length is to say that ξ is a measure of the distance over which the gap parameter, Δ, can vary, for instance in a spatially-varying magnetic field or near a superconductor-normal metal boundary. We also define the intrinsic or BCS coherence length ξ_0 , which is related to the Fermi velocity v_F and the 'energy gap' Δ, by the following relation:

$$\xi_0 = \frac{\hbar v_F}{\pi \Delta} .$$
(2.20)

Experimental measurements of the energy gap (by tunneling or optical absorption) enable one to estimate ξ_0.

Using order of magnitude values for v_F and Δ as listed in Tables 2.1 and 2.2 one can easily estimate the coherence length. However, from Table 2.2 we see that $\xi = 16000$ Å in pure Al, $\xi = 380$ Å in pure Nb, but only about 10 Å in the new superconductors, high-T_c oxides.

Table 2.2: Critical temperature T_c, upper critical magnetic field B_{c2}, magnetic penetration length λ and coherence length ξ for selected superconductors.

	T_c/K	B_{c2}/T	λ/Å	ξ/Å
Al	1.1	0.02*	500	16000
Nb	9.2	0.2	400	380
Nb-Ti	9.5	14	600	450
Nb_3Sn	18.3	24	800	35
Rb_3C_{60}	29.3	~ 50	1600	~ 20
$La_{1-x}Sr_xCuO_4$	38	~ 65	2500	~ 15
$YBa_2Cu_3O_7$	92	~ 120	4000	~ 10

*Al exhibits type-I superconductivity, so the thermodynamic critical field B_c is given.

2.7.3. The Ginzburg-Landau coefficient

The ratio of two characteristic lengths, which we just defined, is called the Ginzburg-Landau ratio κ:

$$\kappa = \frac{\lambda}{\xi} \ . \tag{2.21}$$

It is an important parameter that characterizes the superconducting material. Close to T_c this parameter is independent of temperature and it allows one to distinguish between type-I and type-II superconductors.

If $\kappa < 0.7$ one is dealing with type-I superconductor and if $\kappa > 0.7$ one has a type-II superconductor. The exact critical value of κ that separates type-I from type-II behaviors is $1/\sqrt{2}$ (~ 0.7).

In the latter case the magnetic flux does penetrate into the sample in the form of cylindrical tubes called vortices. Vortices have a radius λ and destroy superconductivity locally within a cylinder of radius ξ. As we shall see in Chapters 3 and 4 it is energetically favorable for type-II superconductors to let the flux penetrate partially in the form of vortices.

2.8. Flux Quantization

The term flux 'quantization' refers to the fact that the magnetic flux threading a superconducting loop cannot have an arbitrary value; it has to be a multiple of $\Phi_0 = \frac{h}{2e}$. In Eq.(2.7) we have introduced the wavefunction $\psi(r)$, to describe the Cooper pairs. In an isolated bulk superconductor, in the absence of an applied magnetic field the phase is the same everywhere. Hence, one can say that there is phase coherence in the whole sample. The absolute value of the phase has no physical meaning; we had mentioned that the gradient of the phase is related to the supercurrent. We now describe a very important consequence of the postulate that all Cooper pairs are described by Eq.(2.7) and that phase coherence extends over the entire sample. In a superconducting ring, the wavefunction has to go through an integral number of oscillations around the loop. The integral number of oscillations explains why magnetic flux inside the ring is quantized; the two are directly related: one gets one quantum of magnetic flux $h/2e$ for each oscillation of the wavefunction.

Consider a ring made of a wire whose diameter is much larger than 2λ. It means that in the presence of an external magnetic field, the magnetic induction and the current, $\mathbf{B}$ and $\mathbf{J}$ respectively, are both equal to zero deep inside the superconducting wire, i.e., $\mathbf{B} = \mathbf{J} = 0$ at a distance greater than λ from the surface.

The superconducting state can be described by a 'macroscopic' wavefunction:

$$\psi(\mathbf{r}) = |\psi(\mathbf{r})| \exp[i\varphi(\mathbf{r})] \ , \quad \psi^*(\mathbf{r}) = |\psi(\mathbf{r})| \exp[-i\varphi(\mathbf{r})] \ , \qquad (2.22)$$

where $\varphi(\mathbf{r})$ is the spatially varying phase of the wavefunction. The momentum (and therefore the velocity) of Cooper pairs is related to the gradient of the phase. More precisely, we have

$$\mathbf{v} = \frac{1}{m} \hbar \nabla \varphi \ .$$

In the presence of an external magnetic field, represented by a vector potential $\mathbf{A}$, this relation reads:

$$\mathbf{v} = \frac{1}{m} (\hbar \nabla \varphi - 2e\mathbf{A}) \ . \qquad (2.23)$$

Indeed, if we have a charge q and we suddenly apply an external magnetic field, described by the vector potential $\mathbf{A}$, we shall get an electric field

$$\mathbf{E} = -\frac{d\mathbf{A}}{dt} \ .$$

This electric field gives a momentum to the particle of $-q\mathbf{A}$. This is the origin of the expression Eq.(2.23) for the velocity.

The current is defined as

$$\mathbf{J} = 2e \, n_s \mathbf{v}_s \qquad (2.24)$$

and can be written as

$$\mathbf{J} = \frac{2e}{m} |\psi|^2 (\hbar \nabla \varphi - 2e\mathbf{A}) \ . \qquad (2.25)$$

Inside the sample

$$\mathbf{J} = 0 \ , \quad \text{i.e.,} \quad \hbar \nabla \varphi = 2e\mathbf{A} \ . \qquad (2.26)$$

Now let us take any path C around the interior of the ring deep inside the sample (see Figure 2.10). At any point, Eq.(2.26) holds, so we have

$$\oint \hbar \nabla \varphi dl = \oint 2e\mathbf{A} dl \ . \qquad (2.27)$$

As φ is a phase, we can increase it by any multiple of 2π without altering ψ. If we go around the path C, the phase change is a multiple of 2π:

$$\oint \nabla\varphi dl = 2\pi n \ , \tag{2.28}$$

where n is an integer. As is well known from electrodynamics, the circulation along the path of $\mathbf{A}$ is the flux enclosed by C, so we have

$$\hbar 2\pi n = 2e\Phi \quad \Rightarrow \quad \Phi = n\frac{h}{2e} \ , \tag{2.29}$$

i.e., the flux inside the ring is an integral number of the flux quantum; h is the Planck constant. The flux quantum is defined as

$$\Phi_0 = \frac{h}{2e} = 2.0678 \times 10^{-15} \ \text{Weber} \ . \tag{2.30}$$

As we shall see in Chapter 6, superconducting devices can measure this tiny variation of magnetic flux which is exceedingly important in metrology and advanced instrumentation and, consequently, in fundamental physics.

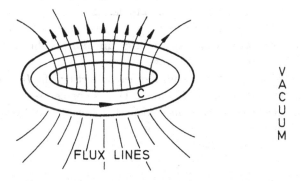

Figure 2.10: Magnetic flux through a superconducting ring.

The first demonstration of flux quantization was performed on superconducting cylinders, which have thin walls and were formed by evaporating a superconductor onto the surface of a glass fiber. The cylinder was cooled through its critical temperature in an external field, and the magnetic moment associated with the trapped flux was measured by vibrating the sample and observing the induced voltage in a nearby coil.

The flux threading the ring, Φ, is the sum of the external flux, Φ_{ext}, and the flux due to supercurrents, Φ_s:

$$\Phi = \Phi_{\text{ext}} + \Phi_s \ . \tag{2.31}$$

As Φ_{ext} is not quantized, it means that the supercurrents adjust themselves in order to satisfy the condition that Φ is an integral number of Φ_0. Flux quantization is one of the most compelling pieces of evidence for the validity of the description of the superconducting state in terms of the macroscopic wavefunction.

2.9. Josephson Effects

2.9.1. Single-particle tunneling

Let us consider two metals (in the normal state) separated by a very thin insulating barrier layer: for example, two electrodes of Al metal separated by a thin film of Al_2O_3 (Figure 2.11). If the insulating layer is thick (say > 100 Å) the conduction electrons cannot propagate through such a barrier. However, if it is sufficiently thin (~ 10–20 Å) there is a significant probability that an electron will tunnel through the barrier. This is the quantum effect called **tunneling** and is well understood and explained in all textbooks of quantum mechanics.

For normal metals and low voltages the *I-V* curve (current-voltage relation) of our sandwich structure (often called **tunneling junction**) is ohmic: the current is directly proportional to the applied voltage.

However, if one of the metals becomes superconducting, the *I-V* curve changes into a characteristic form drawn in Figure 2.11). As one can see from this figure, at absolute zero no current can flow until some characteristic voltage V_g is applied. V_g is directly proportional to the energy gap:

$$V_g = \frac{E_g}{2e} = \frac{\Delta}{e} \ . \tag{2.32}$$

Obviously, at finite temperatures the thermal excitation of electrons allows the passage of a very small current through the barrier even at low voltages.

2.9.2. dc Josephson effect

Using microscopic BCS theory, Brian Josephson predicted in 1962 that if two superconductors are separated by a sufficiently thin layer of insulator (~ 10 Å thick insulating oxide layer 'sandwiched' between two Nb layers, for example) weak currents of Cooper pairs can tunnel through the potential barrier *without any applied voltage*, i.e., the resistance is zero. This is the so-called dc Josephson effect.

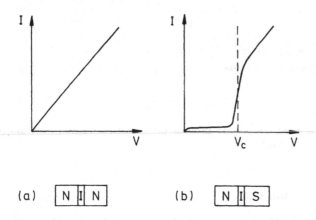

Figure 2.11: a) Linear current-voltage characteristics for two normal metals separated by a thin insulating barrier. b) The *I-V* characteristics when a superconductor replaces one of the metals of diagram (a).

To understand this phenomenon, consider two separate superconductors, each described by its wavefunction. These wavefunctions are not related. If we now couple these two superconductors, their wavefunctions become locked together and the phase coherence extends through both of them. A phase difference can exist between both superconductors which means that a superconducting current can flow.

Consider now two identical superconducting electrodes, A and B, separated by a thin insulating barrier (as schematically presented in Figure 2.11). For sufficiently weak supercurrents the Cooper pairs will tunnel through the barrier and the Josephson junction will behave like a superconductor — there is no voltage difference across the barrier. At zero voltage the current that passes through the junction depends on the phase difference $\varphi_A - \varphi_B$ of the macroscopic wavefunction, Eq.(2.22) of the ensemble of Cooper pairs in superconducting electrodes A and B:

$$I = I_0 \sin \gamma \, , \tag{2.33}$$

where $\gamma = \varphi_A - \varphi_B$, is the phase difference across the junction.

If the injected current I_g through the junction is higher than the critical value I_0 the junction becomes resistive and there appears a voltage difference V across the junction. One can therefore define a characteristic voltage drop V_t across the junction:

$$V_t = RI_0 \; , \qquad (2.34)$$

where R is the resistance of the junction.

Theoretical considerations give

$$V_t = \frac{\pi\Delta}{2e} \; . \qquad (2.35)$$

For metals most commonly used as electrodes, lead and niobium, this voltage is of the order ~ 2.5 mV.

2.9.3. ac Josephson effect

Analyzing the behavior of the junction when a voltage V is applied across it, Josephson has predicted that there will be a flow of an ac current. Moreover, he showed that V is related to the time variation of the phase difference γ by the following relation:

$$\frac{d\gamma}{dt} = \frac{2e}{\hbar}V \; . \qquad (2.36)$$

In such a case one observes characteristic oscillations of the supercurrent whose frequency is given by the relation

$$\nu = \frac{2e}{h}V = \frac{V}{\Phi_0} \; . \qquad (2.37)$$

The ratio ν/V is given in terms of fundamental constants and corresponds to 484 MHz/μV ($\lambda/V = 620$ μm/μV).

This is the so-called **ac Josephson effect**.

Both effects were confirmed experimentally and, as we shall see in Chapters 6 and 8, they represent the basis for contemporary Josephson device technology.

Summary

1. The scattering of 'free' conduction electrons by imperfections gives rise to resistivity in a normal metal at low temperatures. In a superconductor below T_c a quantum fluid of electron pairs is formed. There is **no scattering** of 'individual' pairs of the coherent fluid and therefore **no resistivity**.

2. The superconducting state (i.e., the ensemble of all Cooper pairs) is described by a macroscopic wavefunction:

$$\Psi(\mathbf{r}) = |\Psi(\mathbf{r})|e^{i\varphi(\mathbf{r})} .$$

The square modulus, $\Psi(\mathbf{r})\Psi^*(\mathbf{r})$ is related to the number of Cooper pairs, $n_s(\mathbf{r})$, and the phase gradient, $\nabla\varphi(\mathbf{r})$, is related to the superfluid current

$$\mathbf{J} = \frac{2e}{m}|\Psi(\mathbf{r})|^2(\hbar\nabla\varphi - 2e\mathbf{A}) ,$$

where $\mathbf{A}(r)$ is the vector potential.

3. Materials that always expel completely the flux before becoming normal are called **type-I superconductors**. All pure superconducting metals are type-I except for niobium and vanadium.

For **type-II superconductors** there are two critical fields: a lower one B_{c1} and the upper one B_{c2}. The flux is completely expelled only up to the field B_{c1}. Between B_{c1} and B_{c2}, the superconductor is said to be in the *mixed* state. The Meissner effect is only partial. Above B_{c2} the material returns to the normal state.

4. The gap parameter, which is the characteristic energy of the superconductor, is related to the critical temperature by the BCS relation:

$$E_g = 2\Delta = 3.5k_BT_c .$$

For all frequencies much higher than the frequency that corresponds to the energy gap

$$E_g = h\nu ,$$

where ν is the frequency in Hz, the electromagnetic response in the superconducting state is identical to the response of the normal state. The change in the frequency response occurs at $\nu \sim 10^{11}$ and 10^{12} Hz in the conventional and high-T_c oxides respectively.

5. The characteristic lengths of a superconductor are:

i) the **penetration depth** λ of the order ~ 100–1000 Å in conventional *type-II superconductors*,

ii) the **coherence length** ξ of the order ~ 100 in conventional type-II superconductors.

6. Magnetic flux through a superconducting ring is quantized in units of the **flux quantum**, $\Phi_0 = \frac{h}{2e} = 2.0678 \times 10^{-15}$ Weber.

7. At zero voltage the current that passes through the superconductor-insulator-superconductor junction depends on the phase difference γ of the 'macroscopic' wavefunction (2.22) of Cooper pairs between superconducting electrodes A and B:

$$I = I_0 \sin \gamma \ ,$$

where $\gamma = \varphi_A - \varphi_B$, is the phase difference across the junction. This is the **continuous (or dc) Josephson effect**.

8. When an external voltage V is applied across the junction, one observes characteristic oscillations of the supercurrent whose frequency is given by the relation

$$\nu = \frac{2e}{h}V = \frac{V}{\Phi_0} \ .$$

The ratio ν/V is 484 MHz/μV; $1/V = 620$ μm/μV.

This is the **ac Josephson effect**.

Further Reading

N. W. Ashcroft and N. D. Mermin: *Solid State Physics*, Holt-Saundres International Editions, 1976

J. P. Burger: *Supraconductivité, des Metaux, des Alliages et des Films Minces*, Masson, Paris, 1974

R. P. Feynman, R. B. Leighton, and M. Sands: *The Feynman Lectures On Physics*, see Chapter 21 in Vol. III, Addison-Wesley, 1966

C. Kittel: *Introduction to Solid State Physics*, John Wiley, New York, 1986

A. C. Rose-Innes and E. H. Rhoderick: *Introduction to Superconductivity*, Pergamon Press, Oxford, 1969

Chapter 3.   ELEMENTARY PHENOMENOLOGICAL THEORY

Preview

In the first part of this chapter we present a simple London model which successfully describes the electrodynamics of superconductors. However, the major portion of this chapter is devoted to a presentation of the Ginzburg-Landau theory; we elucidate numerous experimental facts discussed in Chapters 1 and 2. We emphasize several fundamental results which are essential for comprehending a variety of properties of different superconducting materials.

3.1. Introduction

The microscopic theory of superconductivity was formulated by Bardeen, Cooper and Schrieffer (BCS theory) in 1957, almost five decades after the discovery of the phenomenon in 1911 by Kamerlingh Onnes. It is an elegant but mathematically complex theory that cannot be adequately treated at the elementary level. In this Chapter we therefore present a phenomenological, macroscopic description of superconductivity and in Chapter 5 briefly discuss the results of the microscopic theory.

One of the simplest ideas which was put forward in 1934 for describing superconductivity was the two-fluid model. Some properties can be understood with the simple assumption that some electrons behave in the normal way as nearly free electrons, while others exhibit anomalous behavior. Developing this idea, F. and H. London were able to describe the electrodynamics of what is now called type-II superconductors. In those days it was a major step towards an understanding of superconductivity; nowadays the London theory can easily be deduced from the microscopic theory. Because Londons' approach is very useful as an introductory theoretical consideration we present it in the first section.

The phenomenological theory (sometimes called 'macroscopic' or 'quasi-macroscopic') of the superconducting transition was developed jointly by Ginzburg and Landau (GL theory) in 1950. Their theory, although originally phenomenological, proved to be exact and very powerful. As we shall see, it also enables one to develop an insight and phenomenological understanding of high-T_c superconducting oxides.

3.2. The London Model

F. and H. London started with the idea that one has to modify the usual electrodynamic equations in order to describe the Meissner effect; of course Maxwell equations always remain valid. Thus, it is Ohm's law that has to be modified. In order to do that, they used a two-fluid model. Of the total density n of electrons, there is a fraction n_s that behaves in an abnormal way and represents superconducting electrons. These are not scattered by either impurities or phonons, thus they do not contribute to the resistivity. They are freely accelerated by an electric field. If $\mathbf{v}_s$ is their velocity, the equation of motion can be written as

$$m\frac{d\mathbf{v}_s}{dt} = e\mathbf{E} \ . \tag{3.1}$$

We can now define a superconducting current density

$$\mathbf{J} = n_s e \mathbf{v}_s \tag{3.2}$$

which obeys the following equation:

$$\frac{d\mathbf{J}}{dt} = \frac{n_s e^2}{m}\mathbf{E} \ . \tag{3.3}$$

Using the Maxwell equation, curl $\mathbf{E} = \mu_0 \frac{\partial \mathbf{h}}{\partial t}$ in which we replaced magnetic induction $\mathbf{B}$, which varies on the macroscopic scale, by the local microscopic field $\mathbf{h}$ ($\mathbf{B}$ is an average of microscopic $\mu_0 \mathbf{h}$) we obtain:

$$\frac{\partial}{\partial t}\left(\text{curl } \mathbf{J} + \frac{\mu_0 n_s e^2}{m}\mathbf{h}\right) = 0 \ . \tag{3.4}$$

F. and H. London noticed that with Ohm's law and an infinite conductivity, Eq. (3.4) leads to $\partial \mathbf{h}/\partial t = 0$. An infinite conductivity only implies that the magnetic field cannot change, which is contrary to the experimental evidence. Thus they integrated equation (3.4) and took the following particular solution:

$$\text{curl } \mathbf{J} + \frac{\mu_0 n_s e^2}{m}\mathbf{h} = 0 \ . \tag{3.5}$$

This is the London equation which describes the electrodynamics of a superconductor. We now show that it leads to the flux expulsion, the definition of the penetration depth, and a relation between the supercurrent $\mathbf{J}$ and the vector potential $\mathbf{A}$. In order to show how it leads to the Meissner effect, we take the Maxwell equation

$$\mathbf{J} = \text{curl } \mathbf{h} \ . \tag{3.6}$$

By applying the curl-operator to both sides and combining with Eq. (3.5), we obtain

$$\text{curl curl } \mathbf{h} + \mu_0 \frac{n_s e^2}{m}\mathbf{h} = 0 \ , \tag{3.7}$$

or

$$-\Delta \mathbf{h} + \mu_0 \frac{n_s e^2}{m} \mathbf{h} = 0 \ . \tag{3.8}$$

This equation enables one to calculate the local field inside the superconductor and it is another expression of the London equation. Below we show a simple example of the Meissner effect.

Solution for a planar superconductor/vacuum interface

Writing Eq. (3.8) for a one-dimensional problem, we get

$$\frac{d^2 h}{dx^2} = \frac{h}{\lambda^2} \tag{3.9}$$

where we define

$$\lambda^2 = \frac{m}{\mu_0 n_s e^2} \ . \tag{3.10}$$

If we consider a uniform, infinite superconductor in the region $x > 0$ and apply the magnetic field H_0 parallel to the surface, the field inside the superconductor is given by the solution of this equation:

$$h_0 = H \exp(-x/\lambda) \ . \tag{3.11}$$

The field vanishes in the interior of the superconductor (Figure 3.1).

λ is the *London penetration depth* that measures the extension of the penetration of the magnetic field inside the superconductor. It shows that, in order to have zero field within the bulk of the material, one must have a sheet of superconducting current which flows within λ from the surface and which creates an opposite field inside the superconductor that cancels the externally applied magnetic field. Therefore Eq. (3.8) describes well the Meissner effect.

We now give the relation between the supercurrent $\mathbf{J}$ and the vector potential $\mathbf{A}$. As $\mu_0 \mathbf{h} = \text{curl } \mathbf{A}$, we get from London equation (3.5)

$$\mathbf{J}(\mathbf{r}) = -\frac{1}{\mu_0 \lambda^2} \mathbf{A}(\mathbf{r}) \tag{3.12}$$

in the gauge div $\mathbf{A} = 0$. Equation (3.12) replaces Ohm's law for a superconductor. Equations (3.5), (3.8) and (3.12) are different ways of writing the London equation for type-II superconductors.

Additional remarks

      * To obtain the estimate of order of magnitude of λ we take one electron per site. If a_0 is the interatomic distance, we get for the density of superconducting

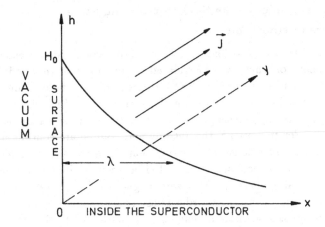

Figure 3.1: Schematic diagram of the penetration of the magnetic field h inside the superconductor according to the London equation.

electrons: $n_s \sim a_0^{-3} \sim 10^{29}$ m^{-3}. Inserting this value into Eq. (3.10) we obtain λ typically of the order of 150 Å. We notice that in general this length is considerably longer than the interatomic distance.

** From Eq. (3.12), it is clear that the London electrodynamics is a local one. It means that the current at some point is given by the vector potential at the same point. In general this is not true and the current at point $\mathbf{r}$ depends on the vector potential in the region around $\mathbf{r}$. The local electrodynamics is only valid for type-II superconductors. For type-I, it is only valid close to T_c. In general the electrodynamics is non-local and Pippard has given a relationship between $\mathbf{J}(\mathbf{r})$ and the value of $\mathbf{A}(\mathbf{r})$ in a small volume around $\mathbf{r}$. The extension of this volume is given by the Pippard coherence length ξ_p, so one has

$$\frac{1}{\xi_p} = \frac{1}{\xi_0} + \frac{1}{l_e} \, , \qquad (3.13)$$

where ξ_0 is the BCS coherence length that we already discussed (in Chapter 2) and l_e is the mean free path of an electron in the metal. As we will focus our attention on type-II superconductors, we shall not need the non-local electrodynamics.

*** In London approach the variation of n_s with temperature is not given. A more general theory is required like, for example, Ginzburg-Landau theory which we shall discuss in the remainder of this chapter.

3.3. Thermodynamics of the Superconducting State

3.3.1. Nature of transition

Superconducting transition belongs to the class of second order phase transitions, the general theory of which was formulated by Landau. In thermodynamics the phase transition is said to be of the first order if there is a discontinuity in volume and entropy as, for example, in the liquid-gas transition. In such transitions one always encounters the hysteresis phenomenon equivalent to supercooling in liquids.

In second order transitions, the entropy and volume are continuous functions and there is no abrupt change at the transition. The transition from the normal (metallic) state to the superconducting state is a second order phase transition that occurs at the critical temperature T_c in zero magnetic field. Therefore one can apply Landau's general theory of second order phase transitions.

General principles of thermodynamics can be applied to the superconducting state. We shall use such an approach to show that the superconducting transition at T_c is of the second order. In order to study the effect of the magnetic field, we shall use the Gibbs potential $G(T, H)$. We calculate the critical field B_c of type-I superconductor by requiring the condition that the Gibbs potentials of the normal (G_n) and the superconducting state (G_s) are equal:

$$G_n(T, H_c) = G_s(T, H_c) \ . \tag{3.14}$$

As $dG = -\mu_0 M dH$, where M is the magnetization and μ_0 is the permeability of vacuum

$$G(T, H_c) = G(T, 0) - \int_0^{H_c} \mu_0 M dH \ . \tag{3.15}$$

In the superconducting phase, $M = -H$, so we have

$$G_s(T, H_c) = G_s(T, 0) + \frac{\mu_0 H_c^2}{2} \ . \tag{3.16}$$

In the normal state, $M = \chi H$, where χ is the magnetic susceptibility. As χ is very small (e.g. $\sim 10^{-5}$) one can write

$$G_n(T, H_c) = G_n(T, 0) \ , \tag{3.17}$$

and hence

$$G_n(T, 0) - G_s(T, 0) = \frac{1}{2}\mu_0 H_c^2 \ . \tag{3.18}$$

As the critical field of type-I superconductors is very small, e.g. typically 10^{-2} Tesla, this energy is very small.

For type-I superconductors it has been experimentally established that the temperature dependence of the critical field H_c is well described within an error of a few percent by the relation

$$H_c(T) = H_o \left[1 - \left(\frac{T}{T_c} \right)^2 \right] . \tag{3.19}$$

The critical field vanishes at the critical temperature.

The relation (3.18) has been derived for a type-I superconductor. However, we will use it as the definition of the thermodynamic critical field of a type-II superconductor. Later we will show that this field also vanishes at T_c for a type-II superconductor.

Using the relation for the entropy: $S = - \left(\frac{\partial G}{\partial T} \right)_H$, one obtains

$$S_n(T,0) - S_s(T,0) = \mu_0 H_c \frac{dH_c}{dT} . \tag{3.20}$$

If $T \to T_c$ as $H_c \to 0$, we have $S_n(T_c) = S_s(T_c)$. The entropy is *continuous* at T_c, hence the transition is a second order phase transition.

By taking the derivative of the entropy, one obtains the expression for the specific heat:

$$C_s(T) - C_n(T) = \mu_0 T \left[\left(\frac{dH_c}{dT} \right)^2 + H_c \frac{d^2 H_c}{dT^2} \right] , \tag{3.21}$$

or, at the critical temperature T_c

$$C_s(T_c) - C_n(T_c) = \mu_0 T_c \left(\frac{dH_c}{dT} \right)^2_{T=T_c} > 0 . \tag{3.22}$$

There is a discontinuity in the specific heat at the transition temperature T_c (see Figure 2.6). This discontinuity is related to the variation of the critical thermodynamic field as a function of temperature at T_c.

3.3.2. Free energy in the London model

We will now show how the London equation can be obtained by minimizing a free energy functional; it will be of use for our later considerations. In a superconductor subjected to an applied magnetic field there are superconducting currents $\mathbf{J}(\mathbf{r})$ and a local field $\mathbf{h}(\mathbf{r})$. The free energy is the sum of the free energy of electrons in the superconducting state without current and the kinetic energy of the current, E_{kin}, and the magnetic energy, E_{mag}. Thus we can write

$$F = \int_V F_s(\mathbf{r}) d^3\mathbf{r} + E_{kin} + E_{mag} \ . \tag{3.23}$$

The kinetic energy can be written as

$$E_{kin} = \int d^3\mathbf{r} \frac{1}{2} m \mathbf{v}_s(\mathbf{r}) n_s \ , \tag{3.24}$$

if n_s is the density of superconducting electrons and $\mathbf{v}_s(\mathbf{r})$ is their velocity at a point $\mathbf{r}$. Using the definition of the superconducting current and the London penetration depth, Eq. (3.10), we can write

$$E_{kin} = \frac{\mu_0}{2} \int d^3\mathbf{r} \lambda^2 \mathbf{J}^2(\mathbf{r}) \ . \tag{3.25}$$

Thus

$$\begin{aligned}
F &= \int d^3\mathbf{r} \left[F_s(\mathbf{r}) + \frac{\mu_0}{2} \lambda^2 \mathbf{J}^2(\mathbf{r}) + \frac{\mu_0}{2} \mathbf{h}^2(\mathbf{r}) \right] \\
&= F_0 + \frac{\mu_0}{2} \int d^3\mathbf{r} \left[\mathbf{h}^2(\mathbf{r}) + \lambda^2 (\text{curl } \mathbf{h}(\mathbf{r}))^2 \right] \ .
\end{aligned} \tag{3.26}$$

If we try to minimize the energy with respect to the field distribution, i.e., if we change $\mathbf{h}(\mathbf{r})$ to $\mathbf{h}(\mathbf{r}) + \delta\mathbf{h}(\mathbf{r})$ we get

$$\delta F = \mu_0 \int d^3\mathbf{r}[\mathbf{h}(\mathbf{r}) + \lambda^2 \text{curl curl } \mathbf{h}(\mathbf{r})]\delta\mathbf{h}(\mathbf{r}) \ .$$

The distribution $\mathbf{h}(\mathbf{r})$ of the field which minimizes the free energy is thus given by the equation

$$\mathbf{h}(\mathbf{r}) + \lambda^2 \text{curl curl } \mathbf{h}(\mathbf{r}) = 0 \ , \tag{3.27}$$

i.e., by the London equation (3.7). The London equation gives the distribution of the field which minimizes the sum of the kinetic energy of the supercurrent and the magnetic energy if the number of superconducting electrons does not vary within the materials.

3.4. The Ginzburg-Landau Theory

The London equation is not applicable to situations in which the number of superconducting electrons, n_s, varies; it does not link n_s with the applied field or current. Therefore we need a more general framework which relates n_s to the external parameters. This is the approach of the Ginzburg-Landau theory which uses the general (Landau) theory of second order phase transitions. Ginzburg and Landau first derive two equations which can be used to calculate both the distribution of the fields and the variation of the number of superconducting electrons. In order to do this they introduce two unknown parameters, α and β. As we will calculate the characteristic quantities of the materials: λ, ξ or H_c, α and β can be eliminated and replaced by two of the measured characteristic quantities (λ, ξ or H_c). One of the greatest successes of this theory was the prediction of the existence of type-II superconductors.

3.4.1. Second-order phase transitions (the Landau theory)

The simplest example of a second order phase transition is the transition from ferromagnetic to paramagnetic state. The spontaneous magnetization M of the sample, which exists below the critical Curie temperature T_{CM}, vanishes above T_{CM}. The magnetization is continuous at the transition and is small, close to the critical temperature. One can easily describe the thermodynamics of such a system by expanding the Helmholtz free energy F in powers of magnetization M, close to the transition point (Figure 3.2):

$$F(T, M) = F(T, 0) + a(T - T_{CM})M^2 + bM^4 + c|\nabla M|^2 \ ,$$

where $\nabla M \equiv \mathbf{grad}\ M$.

By minimizing this energy with respect to M one obtains

$$M = 0 \ \text{ for } \ T > T_{CM} \ ,$$
$$M \neq 0 \ \text{ for } \ T < T_{CM} \ .$$

Using this analysis as the starting point one can describe the magnetic transition. Landau considered that any second order phase transition can be described in the same manner: the magnetization, M, being replaced by another quantity, called order parameter, which is zero above the transition temperature and non-zero below it. As we shall see, this framework inspired Ginzburg and Landau to develop

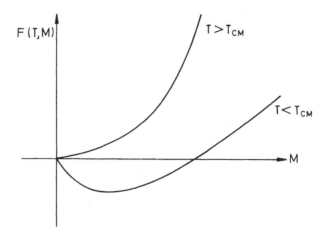

Figure 3.2: Variation of the Helmholtz free energy for two different temperatures: $T < T_{CM}$ and $T > T_{CM}$.

a simple and exact description of superconducting properties near the critical temperature T_c.

3.4.2. Ginzburg-Landau free energy

The basis of this description is the intuitive idea that a superconductor contains superconducting electrons with density n_s and non-superconducting electrons with density $n - n_s$, where n is the total density of electrons in the metal.

One possibility was to use n_s instead of the magnetization M in the Landau theory. However, Ginzburg and Landau have chosen to use a kind of a wavefunction, $\psi(\mathbf{r})$, to describe the superconducting electrons. This function is a complex scalar, Eq. (2.7):

$$\psi(\mathbf{r}) = |\psi\mathbf{r}|e^{i\varphi(\mathbf{r})}$$

and is called the *order parameter*. It has the following properties:

i) Its modulus $|\psi^*\psi|$ can roughly be interpreted as the number of superconducting electrons n_s at a point $\mathbf{r}$.

ii) As in quantum mechanics, the phase $\varphi(\mathbf{r})$ is related to the supercurrent that flows through the material below T_c.

iii) $\psi \neq 0$ in the superconducting state, but zero in the normal state.

Furthermore, Ginzburg and Landau have used the following form of the Helmholtz function:

$$F_s(\mathbf{r}, T) = F_n(\mathbf{r}, T) + \alpha|\psi|^2 + \frac{\beta}{2}|\psi|^4 + \frac{1}{2m}|(-i\hbar\nabla - 2e\mathbf{A})\psi|^2 + \frac{\mu_0\mathbf{h}^2}{2} \qquad (3.28)$$

$$F_s(T) = \int_V F_s(\mathbf{r}, T)d^3r , \qquad (3.29)$$

where s and n denote the superconducting and the normal state correspondingly, while $\hbar = 1.05 \times 10^{-34}$ MKSA is the Planck constant and V is the volume of the sample. In order to see the advantage of using complex functions in describing superconductivity we rewrite Eq. (3.29) using the modulus and phase of the order parameter; hence we get

$$F_s(\mathbf{r}, T) = F_n(\mathbf{r}, T) + \alpha|\psi|^2 + \frac{\beta}{2}|\psi|^4 + \frac{\hbar}{2m}(\nabla|\psi|)^2$$

$$+ \frac{1}{2}|\psi|^2 m\mathbf{v}_s^2 + \frac{\mu_0\mathbf{h}^2(\mathbf{r})}{2} , \qquad (3.30)$$

where we introduced

$$\mathbf{v}_s = \frac{1}{m}(\hbar\nabla\varphi - 2e\mathbf{A}) . \qquad (3.31)$$

One can see that we have obtained the Landau expansion plus the free energy of the field and the current. If the order parameter does not vary in space, one gets back exactly to the London free energy and London equation by carrying out the minimization. If there is no magnetic field and the order parameter has no phase, one obtains the usual Landau theory. The Ginzburg-Landau free energy is thus the way to introduce the London idea in the usual second order phase transition.

Equation (3.28) introduces, *ab initio*, two phenomenological parameters, α and β, into the free energy. The fourth term in Eq. (3.28) is the energy associated with variations of ψ in space. It is written as if ψ represents a true quantum mechanical wavefunction; $\mathbf{A}(\mathbf{r})$ is the vector potential at a point $\mathbf{r}$ and $\mathbf{h}$ is the microscopic field at the same point. As we know from electromagnetism: $\mu_0\mathbf{h} = \text{curl } \mathbf{A}$.

The Helmholtz energy is the integral, over the total volume of the sample, of energy density that depends on the point of consideration. As in the Landau theory one takes

$$\alpha = a(T - T_c),$$

$$\beta = \textbf{positive constant, independent of } T . \qquad (3.32)$$

As we shall see these phenomenological parameters will be determined by fitting the experimental results to the predictions of the Ginzburg-Landau theory.

Some additional remarks

* The coefficient $2e$ in front of the vector potential in Eq. (3.28) (where e is the charge of an electron) was added <u>after</u> the development of the microscopic BCS theory in 1957. Effectively, in 1952 when the Ginzburg-Landau theory was proposed, there was no reason to take the charge as $2e$. At that time it was not clear that superconductivity was linked with the existence of *pairs* of electrons, Cooper pairs, which evidently have charge $2e$.

** $\mathbf{A}(\mathbf{r})$ represents the microscopic vector potential which is an unknown quantity. It is due not only to the applied external field but also to the superfluid current (= supercurrent) that can exist in the interior of the specimen. The theory has to determine $\psi(\mathbf{r})$ and $\mathbf{A}(\mathbf{r})$ which are both unknown functions and which determine all thermodynamic and electrodynamic properties of the superconductor.

3.4.3. Ginzburg-Landau equations

In order to determine the order parameter $\psi(\mathbf{r})$ and the vector-potential $\mathbf{A}(\mathbf{r})$ we minimize the Helmholtz free energy with respect to ψ and $\mathbf{A}$. By this double minimization one gets two equations named after their authors, **Ginzburg-Landau (GL) equations**:

$$\alpha\psi + \beta|\psi|^2\psi + \frac{1}{2m}(i\hbar\nabla - 2e\mathbf{A})^2\psi = 0 , \qquad (3.33a)$$

$$\mathbf{J} = \text{curl } \mathbf{h} = \frac{e}{m}\left[\psi(-i\hbar\nabla - 2e\mathbf{A})\psi + \text{ c.c.}\right] . \qquad (3.33b)$$

These two equations are coupled and should therefore be solved simultaneously. The first gives the order parameter while the second enables one to describe the supercurrent that flows in the superconductor ($\mathbf{J} = \text{curl } \mathbf{h}$). On the basis of these two equations we shall be able to calculate the characteristic quantities which we briefly and phenomenologically discussed in the introductory Chapters 1 and 2.

3.4.4. Consequences of Ginzburg-Landau equations

Thermodynamic critical field

Consider the first equation (3.33a) with no magnetic field. In a homogeneous case one obtains

$$|\psi_0|^2 = -\frac{\alpha}{\beta} \ . \tag{3.34}$$

No solution exists except for $T < T_c$ where $\alpha = a(T - T_c)$ is negative, i.e., superconductivity appears below T_c. Thus

$$F_s(T, 0) - F_n(T, 0) = -\frac{\alpha^2}{2\beta} \ ,$$

and from Eq. (3.18)

$$\mu_0 H_c^2 = \frac{\alpha^2}{\beta} \ . \tag{3.35}$$

Magnetic penetration depth

If we now apply a small magnetic field and assume that we can neglect the variations of ψ, the second Ginzburg-Landau equation (3.33b) gives:

$$\mathbf{J} = \text{curl } \mathbf{h} = -\frac{4e^2}{m}\mathbf{A}|\psi_0|^2 \ .$$

Taking

$$\frac{1}{\lambda^2} = 4e^2\frac{|\psi_0|^2}{m}\mu_0 \ , \tag{3.36}$$

one obtains the London equation. λ is the London penetration depth which we have already defined within the London model, if we put $n_s = 4|\psi_0|^2$. It enables one to calculate the distribution of current and magnetic field:

$$\mu_0\text{curl } \mathbf{h} = -\frac{1}{\lambda^2}\mathbf{A} \ . \tag{3.37}$$

As can be seen from Table 2.2, λ can vary from several thousands of angstroms down to less than a hundred. We remark that λ depends on the temperature; as $\psi_0 \to 0$ at $T_c, \lambda \to \infty$: at the transition the field completely penetrates the sample.

Coherence length

Consider the first Ginzburg-Landau equation (3.33a) in one-dimensional case without external magnetic field:

$$-\frac{\hbar^2}{2m}\frac{d^2\psi}{dx^2} + \alpha\psi + \beta|\psi|^2\psi = 0 \ .$$

This equation defines the length scale:

$$\xi^2(T) = \frac{\hbar^2}{2m|\alpha|} \ .$$

(3.38)

The solution of this equation depends only on $x/\xi(T)$.

This length ξ is called the *coherence length* and represents the length over which the order parameter $\psi(\mathbf{r})$ varies, when one introduces a perturbation at some point. This length also diverges when $T \to T_c$ as $\alpha \to 0$.

Ginzburg-Landau parameter

If both characteristic lengths diverge at T_c in the same manner as $|\alpha|^{-1/2}$, their ratio κ, which is called the Ginzburg-Landau parameter, *does not depend on the temperature*:

$$\kappa = \frac{\lambda}{\xi} \ .$$

(3.39)

Actually κ is the only parameter that really appears in Ginzburg-Landau equations.

3.4.5. Applications of Ginzburg-Landau equations

When one applies Ginzburg-Landau equations to a particular superconductor, the values of α and β can be obtained from experimental measurements.

Expression (3.35) is the first equation that enables one to calculate α and β from experimental data for a given superconducting material. The definition of λ, Eq. (3.36), is the second one; hence by combining the two, we can calculate α and β as functions of H_c and λ:

$$\alpha = -\frac{4e^2}{m}\mu_0{}^2 H_c^2 \lambda^2 \ ,$$

(3.40)

$$\beta = \left(\frac{4e^2}{m}\right)^2 \mu_0^3 H_c^2 \lambda^4 \ .$$

(3.41)

If one measures the length in units of λ and the field in units of H_c one can demonstrate that the free energy of Ginzburg and Landau depends only on κ. The Ginzburg-Landau parameter κ therefore phenomenologically completely characterizes a given superconductor.

Finally, let us note that by combining Eqs. (3.35) with (3.36) and (3.38) one gets an important relationship between characteristic quantities that we discussed in this section:

$$H_c(T)\lambda(T)\xi(T) = \text{constant} = \frac{\hbar}{2e\mu_0\sqrt{2}} = \frac{\Phi_0}{2\pi\mu_0\sqrt{2}} \ . \tag{3.42}$$

3.5. Type-I and Type-II Superconductors

At the critical temperature T_c, the phase transition from normal to superconducting state is always a second order transition in zero magnetic field. For type-I superconductors, at a given temperature below T_c, the transition from superconducting to a normal state is of the first order at B_c, contrary to the type-II superconductors which undergo second order phase transition at B_{c2}. We will now characterize the difference in behavior of the two types of superconductors.

For a second order transition, the order parameter is small. Taking Ginzburg-Landau equation (3.33a) and limiting the analysis to first order in Φ, we have

$$\frac{1}{2m}(-i\hbar\nabla - 2e\mathbf{A})^2\psi = -\alpha\psi \ . \tag{3.43}$$

In this equation $\mathbf{A}$ is the vector potential of the applied magnetic field. The contribution due to the superfluid currents is given by the second equation: as it is of the order of $|\psi|^2$ it can be neglected. Equation (3.43) is a well-known equation in quantum mechanics. It describes the motion of a charged particle $(q = 2e)$ in the magnetic field. The lowest eigenvalue of this Schrödinger equation is

$$E_0 = \frac{1}{2}\hbar\omega_c \tag{3.44}$$

where ω_c is the cyclotron frequency, $\omega_c = 2eB/m$. We have a nonzero solution for ψ in Eq. (3.43) and hence the appearance of the superconducting state, if $B < B_{c2}$ with

$$\frac{e\hbar B_{c2}}{m} = -\alpha \ . \tag{3.45}$$

Using the definition of ξ, Eq. (3.38), we get

$$B_{c2}(T) = \frac{\Phi_0}{2\pi}\frac{1}{\xi^2(T)} \ . \tag{3.46}$$

Combining it with Eq. (3.42) we get

$$B_{c2} = \mu_0 H_{c2} = \mu_0\kappa\sqrt{2}H_c \ . \tag{3.47}$$

One can distinguish two different situations:

* If $\kappa < 1/\sqrt{2}$, we have $B_{c2} < B_c$. By decreasing the field the superconducting state appears at (and below) B_c with total expulsion of the flux (as was illustrated in Figure 1.2): we have a *type-I superconductor*.

** If $\kappa > 1\sqrt{2}$, we have $B_{c2} > B_c$. The superconducting state appears at and below B_{c2}. As the flux expulsion is not complete we have a *type-II superconductor*.

The difference between these two types of behavior can be understood from the following analysis. To achieve $B = 0$ magnetic energy has to be considerable. Type-II superconductors minimize that energy by creating small normal cylinders called vortices in the mixed superconducting state. As one cannot destroy superconductivity, i.e., vary the order parameter ψ over a distance smaller than the coherence length, the smallest cylinder that one can create has a radius ξ (see Figure 3.3). The creation of such a vortex costs energy:

$$(F_s - F_n) \times \text{volume of the cylinder} = \mu_0 \frac{H_c^2}{2\xi^2} d \ ,$$

where d is the length of the vortex. The creation of such a normal cylinder permits the magnetic field to penetrate into the sample.

However, the field will penetrate the inside of the superconductor only over the distance λ. Thus the energy gain is

$$\mu_0 \frac{H^2}{2} \lambda^2 d \ .$$

The energy balance is in favour of the creation of such vortex if $\lambda \gg \xi$ or $\kappa \gg 1$. This explains why magnetic behavior is different for $\kappa \gg 1$ as compared with $\kappa \ll 1$.

Additional remarks

*From Eq. (3.47) one can immediately see that the measurement of the upper critical field B_{c2} gives **directly** the coherence length of type-II superconductors. If $B_{c2} \sim 3$ Tesla one gets $\xi = 100$ Å.

**The same argument which we used for the flux quantization of a ring (Section 2.9.) applies to the vortex: the flux of a vortex is an integer number of flux quanta. We will show that in fact the flux is exactly Φ_0 for each vortex.

***If the cylinders are close together the superconducting material becomes normal. As each cylinder carries a flux Φ_0 it amounts to a magnetic induction of the order of $\frac{\Phi_0}{\pi\xi^2}$, i.e., B_{c2}. Close to the upper critical field the distance between vortices is the coherence length.

****The fact that the tube must transport a flux Φ_0 gives an immediate estimate of the first critical field B_{c1}. Indeed, the flux associated with the cylinder is $\pi\lambda^2 B$ and its value has to be Φ_0.

The minimum B is

$$B_{c1} \approx \frac{\Phi_0}{\pi\lambda^2} . \tag{3.48}$$

This is the field which is needed to nucleate one tube; the exact formula will be derived in the next section.

3.6. The Lower Critical Field

3.6.1. An isolated vortex

Each vortex transports only one quantum of flux, $\Phi_0 = h/2e$. Therefore the number of vortices gradually increases as the field is raised from B_{c1} to B_{c2}. Qualitatively, the upper critical field, B_{c2}, corresponds to the case where the distance between vortices is equal to the radius of the normal core of the vortex. Near B_{c1}, the distance between the vortices is of the order of λ. The vortices repel each other and it is this magnetic pressure that determines the vortex density. As the field increases, the magnetic pressure and the number of vortices increase. The density of vortices per unit surface n is related to B by the important relation

$$B = n\Phi_0 . \tag{3.49}$$

This enables one to calculate the distance between the vortices as a function of B.

In order to study the flux penetration we start with a description of an isolated vortex. We consider the most interesting practical limit where $\xi \ll \lambda$. In that limit, $\kappa \gg 1$, the second Ginzburg-Landau equation takes the simple form (3.27) when the order parameter is constant, e.g. $r > \xi$, namely over all except a small core region of radius ξ. In order to describe the fact that the vortex has a central core of radius ξ (see Figure 3.3) and carries a flux Φ_0 we transform equation (3.27) into

$$\mathbf{h} + \lambda^2 \text{curl curl } \mathbf{h} = \frac{\Phi_0}{\mu_0}\delta_2(\mathbf{r}) , \tag{3.50}$$

where $\delta_2(\mathbf{r})$ describes a singularity in the plane perpendicular to the field direction at $\mathbf{r} = 0$ and imposes a flux Φ_0 per vortex. By solving Eq. (3.50) one finds how the field decreases from the center of the vortex:

$$\mathbf{h}(\mathbf{r}) = \frac{\Phi_0}{2\pi\lambda^2\mu_0}\mathbf{K}_0\left(\frac{\mathbf{r}}{\lambda}\right) , \tag{3.51}$$

where $\mathbf{K}_0$ is a Hankel function given in standard mathematical tables. At large distances the field decreases as $e^{-r/\lambda}$ (see Figure 3.3).

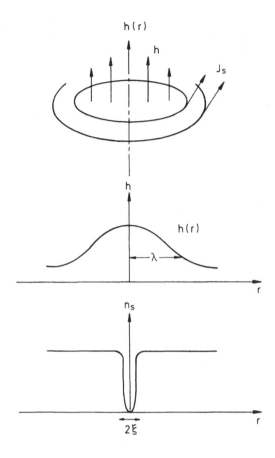

Figure 3.3: The cross section of an isolated vortex in type-II superconductor (after de Gennes 1966).

The superfluid current associated with this field by Eq. (3.6) circulates around the vortex, hence its name. The modulus of this current is

$$|\mathbf{J}| = |\text{curl } \mathbf{h}| = -\frac{d\mathbf{h}}{d\mathbf{r}} ,\qquad (3.52)$$

and it also decays over a distance of the order λ. Figure 3.3 shows also the variation of the square of the order parameter ψ within a vortex, i.e., variation of the number of superconducting electrons. The order parameter vanishes at the core of the vortex; at a distance ξ it attains the value which it has below B_{c1}, i.e., in the absence of vortices. It is easy to calculate the magnetic energy U_v of the vortex, which is the sum of the kinetic energy of the current and the energy of the magnetic field, Eq. (3.26):

$$U_v = \int d^3 r \frac{\mu_0}{2} (h^2 + \lambda^2 J^2) \,, \tag{3.53}$$

i.e.,

$$U_v = \frac{\Phi_0^2}{4\pi \lambda^2 \mu_0} \ln \frac{\lambda}{\xi} \,. \tag{3.54}$$

A vortex can be considered as a whirl of superfluid currents around a normal metal tube.

3.6.2. The lower critical field B_{c1}

The lower critical field B_{c1} corresponds to the thermodynamic limit above which it is energetically favorable for vortices to penetrate into the superconductor. In order to estimate B_{c1}, we calculate the variation of the Gibbs energy which corresponds to the entry of n vortices per unit surface

$$\Delta G = n U_v - B H \,. \tag{3.55}$$

Here we have neglected the interaction between vortices which, close to B_{c1}, are rather far apart. Using Eq. (3.49) we have

$$\Delta G = B \left(\frac{U_v}{\Phi_0} - H \right) \,. \tag{3.56}$$

The vortices will penetrate into the superconductor if this lowers the Gibbs energy. The condition is

$$H \geq H_{c1} = \frac{U_v}{\Phi_0} \,. \tag{3.57}$$

Using (3.54) we finally obtain:

$$B_{c1} = \frac{\Phi_0}{4\pi \lambda^2} \ln \frac{\lambda}{\xi} \,. \tag{3.58}$$

The field B_{c1} is small when $\kappa \gg 1$.

An interesting relation can be obtained for the product $B_{c1} B_{c2}$ if we use Eq. (3.42):

$$B_{c1} B_{c2} = B_c^2 \ln \kappa \,. \tag{3.59}$$

As the value of the thermodynamic critical field does not vary very much in type-II superconductors between 0.1 and 1 T, it means that the higher B_{c2} is, the lower becomes B_{c1}. In high-T_c superconductors where B_{c2} is very high (of the order 100 T) due to the small coherence length, it means that B_{c1} is very small, of the order 10^{-2} Tesla.

3.7. Surface and Interface Effects

3.7.1. Critical field B_{c3}

Our solution of Ginzburg-Landau equations is rigorously valid only for an infinite sample. We have found that below B_{c2} the superconductivity appears within the volume of the sample. Consider now a situation where the sample is semi-infinite in the half-space $x > 0$ and where the field is applied parallel to the surface. To find nonzero solution in this geometry and calculate the critical field we have to solve the same linearized Ginzburg-Landau equation with an additional restriction: that there is no current perpendicular to the surface of the sample. The second Ginzburg-Landau equation yields the supercurrent so at the surface of the sample we impose

$$(-i\hbar\nabla - 2e\mathbf{A})_{\mathbf{n}}\Psi = 0 \ . \tag{3.60}$$

With this boundary condition one can show that the lowest-energy solution of the linearized Ginzburg-Landau equation (3.43) becomes

$$E_0 = 0.59 \times \frac{1}{2}\hbar\omega_c \ .$$

Consequently the critical field at which superconductivity appears is

$$B_{c3} = \frac{1}{0.59}B_{c2} = 1.69 B_{c2} \ . \tag{3.61}$$

This means that if the field B is parallel to the surface and has strength

$$B_{c2} < B < B_{c3} \ ,$$

then a superconducting layer of thickness $\sim \xi$ appears within the surface of the specimen. Of course, if one reduces the field to below B_{c2}, the superconductivity appears in the volume of the sample

The existence of B_{c3} in the aforementioned geometry has a direct consequence on the interpretation of the various experiments. When the magnetic field is parallel to the length of the sample, resistivity measurements will give B_{c3}: the resistivity is zero below that field. The supercurrent passes within the superconducting layer near the surface. If one measures the magnetization, one obtains B_{c2} as this experiment is sensitive to the state of the bulk of the sample and the flux is not being expelled until the field is reduced down to or below B_{c2}.

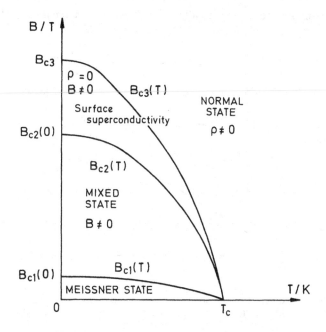

Figure 3.4: Schematic variation of three critical fields of type-II superconductor.

For certain type-I superconductors, the field B_{c3} can be larger than B_c. In that particular case one can observe surface superconductivity.

3.7.2. Proximity effects

Until now we have dealt only with individual superconductors. Now we will consider the effects that result when a normal metal is deposited onto a superconductor. If the electrical contact is of a sufficiently good quality ($\Rightarrow$ interface problem is in most cases a complex materials problem) the normal metal will alter the order parameter ψ close to the interface (see Figure 3.5).

In the Ginzburg-Landau approach, this effect is described by a boundary condition on ψ which implies that the superfluid current cannot escape from the sample; the appropriate boundary condition is a direct generalization of Eq. (3.60). We write for the interface

$$\frac{\partial \psi}{\partial n} - \frac{2ie}{\hbar} A\psi = \frac{1}{b_e}\psi \; , \tag{3.62}$$

where b_e corresponds to a length called the *extrapolation length* (see Figure 3.5).

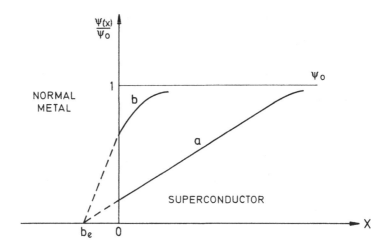

Figure 3.5: Schematic diagram of the order parameter at the interface between a normal metal and a superconductor: a) near T_c, b) at low temperatures (after Deutscher and Müller 1987).

There are several physical effects related to the existence of b_e, like, for example, the decrease of the critical temperature of thin superconducting films deposited on the normal metal or the disappearance of surface superconductivity below B_{c3} under some specific conditions.

The microscopic theory gives for a superconductor-insulator interface

$$b_e \sim \frac{\xi_0^2}{a_0} \; . \tag{3.63a}$$

At a superconductor-normal metal interface in the limit $\xi_0 \ll l_e$ (called 'clean limit') we have:

$$b_e \sim \xi_n = \frac{\hbar v_F}{k_B T} \; , \tag{3.63b}$$

while for $\xi_0 \gg l_e$ (the 'dirty limit') we get:

$$b_e \sim \xi_n = \sqrt{\frac{\hbar v_F l_e}{6\pi k_B T}} \; . \tag{3.63c}$$

In conventional superconductors the coherence length is much larger than the interatomic distance ($\xi_0 \gg a_0$). For superconductors deposited on an insulator, b_e is very large so superconductivity is not affected.

When the coherence length is short, $\xi_0 \sim a_0$, one can expect some important effects. A typical example is the new oxide superconductors: one can expect considerable proximity effects in films or in the bulk that contains some insulating grains.

In the case of a superconductor-normal metal interface, superconductivity can be induced in the normal metal within a sheath of thickness ξ_n. This length is called the *normal coherence length*. The superconductivity is due to Cooper pairs which enter the normal metal.

Summary

1. Local electrodynamics of type-II superconductors is described by the London equation which replaces Ohm's law in a superconductor:

$$\text{curl } \mathbf{J}(\mathbf{r}) + \frac{\mu_0 n_s e^2}{m} \mathbf{h}(\mathbf{r}) = 0 \; ,$$

or in the gauge div $\mathbf{A} = 0$:

$$\mathbf{J}(\mathbf{r}) = -\frac{1}{\mu_0 \lambda^2} \mathbf{A}(\mathbf{r}) \; ,$$

where $\lambda = \sqrt{\frac{m}{\mu_0 n_s e}}$ is the London penetration depth.

2. The phenomenological Ginzburg-Landau theory describes a superconductor close to T_c. The order parameter $\psi(\mathbf{r}) = |\psi| \exp i\varphi(\mathbf{r})$ is a *complex scalar* and $\psi(\mathbf{r})$ is small. The theory introduces *ab initio* two parameters which can be estimated from experimental measurements.

3. While both the characteristic lengths λ and ξ diverge at T_c, their ratio, $\kappa = \lambda/\xi$, which is called the Ginzburg-Landau parameter, does not depend on temperature when close to T_c.

4. If $\kappa < 1/\sqrt{2}$, by decreasing the field the superconducting state appears at (and below) B_c with a total expulsion of the flux: this is the behavior of type-I superconductor.
  For $\kappa > 1/\sqrt{2}$ the superconducting state appears at and below B_{c2}. As flux expulsion is incomplete we have type-II superconductor. The mixed (*vortex*) state exists between B_{c1} and B_{c2}.

5. Measurement of the upper critical field B_{c2} gives directly the coherence length ξ by the relation

$$B_{c2}(T) = \frac{\Phi_0}{2\pi} \frac{1}{\xi^2(T)} \; .$$

6. The lower critical field B_{c1} is given by

$$B_{c1}(T) = \frac{\Phi_0}{4\pi} \frac{1}{\lambda^2(T)} \ln \frac{\lambda}{\xi} \; .$$

7. The relations between the critical fields are

$$B_{c1}B_{c2} = B_c^2 \ln \kappa$$

and

$$B_{c3} = \frac{1}{0.59}B_{c2} = 1.69 B_{c2} \ .$$

8. In the mixed state, the density of vortices n is given by

$$n = \frac{B}{\Phi_0} \ .$$

9. If we have a contact between a normal metal and a superconductor, the order parameter is changed close to the surface and we have to introduce boundary conditions to the Ginzburg-Landau equations. The 'proximity' coherence length in the normal metal is given by

$$\xi_n = \frac{\hbar v_F}{k_B T} \text{ in the 'clean limit' } (l_e \gg \xi_0)$$

and

$$\xi_n = \sqrt{\frac{\hbar v_F l_e}{6\pi k_B T}} \text{ in the 'dirty limit' } (l_e \ll \xi_0)$$

Further Reading

P. G. de Gennes: *Superconductivity of Metals and Alloys*, W.A. Benjamin, New York, 1966

V. L. Ginzburg and D. A. Kirzhnits (editors): *High-Temperature Superconductivity*, Consultants Bureau (Plenum), New York, 1982

E. M. Lifshitz and L. P. Pitaevsky: *Statistical Physics Part 2*, Pergamon Press, 1980

D. Saint-James, G. Sarma and E. Thomas: *Type-II Superconductors*, Pergamon Press, 1969

D. R. Tilley and J. Tilley: *Superfluidity and Superconductivity*, Adam-Hilger, Bristol, 1990

M. Tinkham: *Introduction to Superconductivity*, McGraw-Hill, New York, 1975; reprinted by Robert E. Krieger, Malabar, Florida, 1985

Chapter 4. CRITICAL CURRENT OF TYPE-II SUPERCONDUCTORS

Preview

In this chapter we discuss the behavior of type-II superconductors in an external magnetic field between B_{c1} and B_{c2}. This, so-called, mixed state is required in most applications. We first describe the equilibrium state which is a triangular lattice of flux lines. As no current can flow without dissipation in this ideal state, we discuss metastable states which are mainly caused by inhomogeneities in the material. These metastable states, which lead to irreversible properties, are shown to be the necessary condition to observe non-dissipative supercurrents in these materials. This leads to the concept of critical current and the study of different regimes obtained with a current flowing in the mixed state.

4.1. Mixed State: Stable and Metastable States

If a type-II superconductor is placed in an external magnetic field, $B(B_{c1} < B < B_{c2})$, the flux partially penetrates into the sample. Below B_{c1}, the expulsion is complete so inside the superconductor we have $B = 0$; above B_{c2}, the sample is in the normal state. Between B_{c1} and B_{c2} the flux partially penetrates into the sample in the form of flux tubes (see Figure 4.1), called vortices. As we shall see later, the flux lines form a triangular lattice that one can directly observe.

This mixed state is a stable state and in Chapter 2 we have already seen the magnetization curve for an ideal type-II superconductor. However, what one requires for most applications is the highest possible transport current that can persist even in high magnetic fields. The maximum current that can be transported by type-I superconductors is very small; a relatively weak external magnetic field exceeds the critical field thereby destroying superconductivity. Therefore in most applications one uses type-II superconductors in which one can have high critical currents. However, the current flow creates a magnetic field and vortices in the sample. As we mentioned in Section 2.4 the current flow in the mixed state is a rather complex problem that we shall discuss in this chapter.

First we will show how to obtain the ideal magnetization curve. However, if one makes a measurement of magnetization as a function of field one has difficulties in obtaining the predicted ideal behavior. Samples have to be carefully prepared and thoroughly annealed. But, if one does cold working on such a 'perfect' sample, the behavior becomes completely different (see Figure 4.2). Now the field has difficulties to penetrate due to the inhomogeneities and defects. At B_{c1} the magnetization changes very little; hence this field B_{c1} is difficult to measure. Moreover, when one lowers the field, coming from the mixed state above B_{c1}, the flux has difficulties to be expelled and stays trapped inside. Instead of a reversible magnetization curve one obtains a hysteresis cycle as in a permanent magnet.

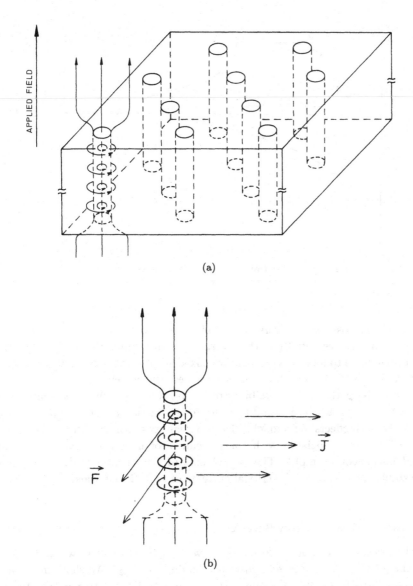

(a)

(b)

Figure 4.1: a) Schematic diagram of the mixed state. Note that vortices form a hexagonal lattice. b) The Lorentz force **F** on a flux line in the presence of the current **J** (adapted from Buckel 1977).

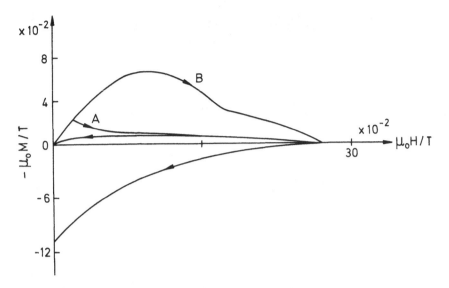

Figure 4.2: An example of the hysteresis in magnetization measurements: curve A represents the hysteresis cycle without annealing; curve B shows the same sample after cold working (simplified diagram, after Livingston 1964).

It is this absence of reversibility which permits the existence of permanent magnets; they are essentially in the metastable state. Similarly, it is the hysteresis that permits one to have a current in the mixed state without dissipation. Indeed if the system is perfect, as we have already noticed, the current will drive the vortices and cause dissipation. The possibility of having zero resistance in the mixed state is directly related to the possibility of hysteresis in the magnetization curve. We will relate both effects in a model called the Bean model or the critical state model. Both effects are associated with the difficulties of having vortex motion within an inhomogeneous sample. The second part of this chapter is devoted to these metastable states, i.e., to the critical currents of type-II materials.

4.2. Interaction between Vortices

Restricting our analysis to materials with high values of κ, we now calculate the energy of two vortices separated by the distance $\mathbf{r}_{12}$. We show that vortices repel each other and that the equilibrium condition for a vortex is that the total supercurrent vanishes at the core of the vortex.

We begin with Eq. (3.27), modified as in Eq. (3.50) by two singularities which describe two vortices: one at the origin and the other at the point $\mathbf{r}_{12}$. As the equation is linear in h, it is easy to show that the field created by the two vortices is just equal to the sum of the fields separately created by each vortex. As a result it can be shown that the total energy, U_t which is given by Eq. (3.53) is a sum of individual energies of the vortices, $2U_v$, plus an interaction term:

$$U_t = 2U_v + U_{12} ,\tag{4.1}$$

with

$$U_{12} = \Phi_0 h(\mathbf{r}_{12}) .\tag{4.2}$$

The interaction energy is equal to ϕ_0 multiplied by $h(\mathbf{r}_{12})$, the field created by one of the vortices and 'sensed' by the other. Since this energy is positive the vortices repel each other. It is the magnetic pressure that forces them to enter the sample and compress themselves when the external field increases. The force exerted by vortex 1 on vortex 2 is given by

$$\mathbf{f}_x = \frac{\partial U_{12}}{\partial x_2} = \Phi_0 J_y .\tag{4.3}$$

The force on vortex 2 is along the line that links the vortices and is proportional to the superfluid current created by vortex 1. As this current is perpendicular to this very link, it is equivalent to the Magnus force of a turbulence in hydrodynamics (Figure 4.1b). In an array of vortices the total force on one vortex vanishes because of the symmetry of the vortex lattice.

In order to study the thermodynamics of an ensemble of vortices in the sample, we have to generalize the Gibbs energy, Eq. (3.56), and take into account the repulsion between vortices

$$G_s(H) - G_s(0) = \frac{BU_v}{\phi_0} + \sum_{ij} U_{ij} - BH .\tag{4.4}$$

By minimizing the Gibbs energy with respect to B, we have

$$H - H_{c1} = \frac{\partial}{\partial B} \sum_{ij} U_{ij} .\tag{4.5}$$

Using this equation, one can calculate the most stable lattice and the magnetization curve. It turns out that the triangular lattice is the most stable one. In that case, the distance a between two vortices is given by

$$B = \frac{2}{\sqrt{3}} \frac{\Phi_0}{a^2} ,$$

or

$$a = 1.072\sqrt{\frac{\Phi_0}{B}} \; . \tag{4.6}$$

The vortex lattice parameter a decreases with increasing magnetic field.

4.3. The Abrikosov Lattice

Penetration of flux in type-II superconductors through a lattice of flux lines has been predicted by Abrikosov. In fact his analysis was for fields close to B_{c2} where our presentation is not valid. Indeed, in that case vortices are very close and the order parameter is reduced everywhere. One has to use the full Ginzburg-Landau equations and calculate both the order parameter and the distribution of the field. Abrikosov's calculation is lengthy so we only quote his results. The average value $< |\psi|^2 >$ of the order parameter as a function of magnetic field vanishes linearly close to B_{c2}:

$$< |\psi|^2 >= |\psi_0|^2 \frac{1}{1.16} \frac{2\kappa^2}{2\kappa^2 - 1} \left(1 - \frac{H}{H_{c2}} \right) \; . \tag{4.7}$$

The free energy is given by

$$F = \frac{B^2}{2\mu_0} - \frac{1}{2\mu_0} \frac{(B_{c2} - B)^2}{1 + (2\kappa^2 - 1)1.16} \; . \tag{4.8}$$

From the definition

$$H = \frac{\partial F}{\partial B} \tag{4.9}$$

and

$$M = \frac{B - \mu_0 H}{\mu_0} \tag{4.10}$$

one can show that

$$M = \frac{H_{c2} - H}{1.16(2\kappa^2 - 1)} \; . \tag{4.11}$$

For a given temperature, the magnetization vanishes at H_{c2} as a linear function of the applied magnetic field. Measurements of the slope of the magnetization vs. magnetic field close to H_{c2} enable one to determine κ. This represents one of the few direct measurements of the Ginzburg-Landau parameter $\kappa = \frac{\lambda}{\xi}$, which is one of the fundamental quantities that characterize a given superconductor.

The flux lattice can be directly visualized by magnetic decoration techniques. One scatters finely divided ferromagnetic particles on the surface of a sample with a magnetic field perpendicular to the surface. These magnetic particles tend to accumulate on the flux line cores where they remain on warming up to room temperature. Subsequent examination under an electron microscope shows clearly the flux line lattice (see Figure 4.3).

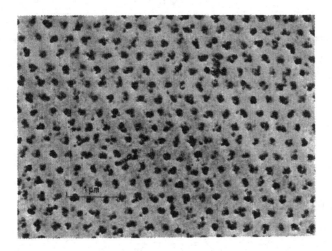

Figure 4.3: Triangular lattice of flux lines (after Essmann and Träuble 1967).

4.4. Anisotropic Type-II Superconductors

So far we have restricted our analysis to isotropic superconductors. Since the new high-temperature oxide superconductors are highly anisotropic, we shall give here, without proof, the value of the lower and upper critical fields. As these new materials have layered structures, one has to replace the electron mass by the mass tensor which has two principal values: m^c along the c-axis and m^{ab} in the ab-plane (see Chapter 7, Figure 7.1).

As the mass along the c-axis is much larger than in the ab-plane, we introduce a small quantity

$$\varepsilon = \sqrt{\frac{m^{ab}}{m^c}} \ll 1 \ .$$

For example, this ratio is of the order 0.2 for the $YBa_2Cu_3O_7$ oxide superconductor. The Ginzburg-Landau equations introduce two coherence lengths:

$$\xi^c = \frac{h}{\sqrt{2m^c\alpha}} \ ,$$

and

$$\xi^{ab} = \frac{h}{\sqrt{2m^{ab}\alpha}} \ , \tag{4.12}$$

hence we have $\xi^c \ll \xi^{ab}$ for the coherence lengths along the c-axis and in the ab-plane, respectively. The order parameter changes over a smaller length along the c-axis than in the ab-plane.

We also introduce two penetration depths :

$$\lambda^c = \sqrt{\frac{m^c}{\mu_0 n_s e^2}}$$

and

$$\lambda^{ab} = \sqrt{\frac{m^{ab}}{\mu_0 n_s e^2}} \ , \tag{4.13}$$

hence in high-T_c oxides $\lambda^c \ll \lambda^{ab}$. We have

$$\varepsilon = \frac{\xi^c}{\xi^{ab}} = \frac{\lambda^{ab}}{\lambda^c} \tag{4.14}$$

λ^c is involved when the supercurrent shielding the field has to flow along the c-axis, λ^{ab} when it flows in the ab-plane.

By solving the anisotropic Ginzburg-Landau equations as a function of the angle θ (the angle between the c-axis and the magnetic field) we get:

$$B_{c2} = \frac{\Phi_0}{2\pi(\xi^{ab})^2}(\cos^2\theta + \varepsilon^2\sin^2\theta)^{-1/2} \ . \tag{4.15}$$

The limiting values of B_{c2} are

$$B_{c2}^c = \frac{\Phi_0}{2\pi(\xi^{ab})^2} \quad \text{for} \quad B\|c \ . \tag{4.16}$$

$$B_{c2}^{ab} = \frac{\Phi_0}{2\pi\xi^{ab}\xi^c} \quad \text{for } B \perp c . \tag{4.17}$$

The coherence lengths involved are the coherence lengths in the plane perpendicular to the field. The critical field parallel to the layer is much higher than the field perpendicular to it. For the ratio we have

$$\frac{B_{c2}^c}{B_{c2}^{ab}} = \frac{\xi^c}{\xi^{ab}} = \varepsilon . \tag{4.18}$$

The triangles of the Abrikosov lattice are no longer equilateral except in the case where the field is parallel to the c-axis. In all other cases one obtains isosceles triangles. For example, if the field is applied along the b-axis, the ratio of the height of the triangle d_a to the distance d_c between vortices along the c-axis is

$$\frac{d_a}{d_c} \sim \frac{\lambda^c}{\lambda^{ab}} .$$

For the first critical field one gets

$$\begin{aligned} B_{c1}^c &= \frac{\Phi_0}{4\pi(\lambda^{ab})^2} \ln \frac{\lambda^{ab}}{\xi^{ab}} \quad \text{for } B\|c , \\ B_{c1}^{ab} &= \frac{\Phi_0}{4\pi\lambda^{ab}\lambda^c} \ln\left(\frac{\lambda^{ab}\lambda^c}{\xi^{ab}\xi^c}\right) \quad \text{for } B\|ab . \end{aligned} \tag{4.19}$$

We have the inequality $B_{c1}^{ab} \ll B_{c1}^c$.

Hence the vortex lines exhibit a trend towards lying in the layer plane where they are formed at relatively small fields. For the angle θ we have

$$B_{c1}^2(\theta)\left[\frac{\cos^2\theta}{\left(B_{c1}^c\right)^2} + \frac{\sin^2\theta}{\left(B_{c1}^{ab}\right)^2}\right] = 1 . \tag{4.20}$$

Further discussion of this topic is given in Chapter 7 (see Section 7.8.2).

4.5. Irreversible Properties: Metastable States

As we have pointed out, it is difficult to obtain the ideal magnetization curve. As soon as the material has defects, we obtain a hysteresis curve. This interesting fact shows that vortices have difficulties moving, either to enter or to leave the sample. This difficulty to move vortices permits the sample to sustain a current without dissipation. Indeed we show that a current drives a vortex perpendicularly to the direction of the current. If the vortex moves it means a local change of flux and, e.g. through Maxwell equations, the appearance of an electric field. If we have both an electric field and the current we have dissipation. In an ideal type-II superconductor, any small current will create dissipation as there is nothing to prevent the motion of the lattice of vortices. The critical current would be the current which creates a field H_{c1} at the surface of the sample, thus a very small current. If d is the dimension of the sample, we have

$$J \sim \frac{H_{c1}}{d} \; . \tag{4.21}$$

Now if we have a strong hysteresis, which means that the vortices have difficulties moving, a current will flow without moving them if the force exerted on the vortices is not strong enough to set them into motion. The superconductor will support a current without dissipation. The dissipation will occur only when the force exerted on the vortices is strong enough to overcome the barrier which prevents vortices to move, i.e., there will be a critical current density J_c for the material. This J_c is of course very dependent on the irreversibility of the sample; it is zero in an ideal sample. We can increase J_c by increasing the irreversibility.

4.5.1. Depairing critical current

If we were able to completely prevent the motion of vortices, what would the critical current be? Infinite or not? In fact, there is an intrinsic critical current that a superconductor can support which is called the depairing critical current. It is the current which destroys pairs and thus superconductivity. This is understandable with the same reasoning that shows the existence of the critical magnetic field. The superconducting state has gained an energy per unit volume, $\mu_0 \frac{H_c^2}{2}$, over the normal state; if we put a superconducting sample in a situation which increases its energy by this amount, it will go back to the normal state. If we introduce currents in the superconducting state, we increase the energy by a value given by the kinetic energy of the current. Let us consider the London model. Using Eq. (3.2) we get for the energy of the current:

$$n_s \frac{1}{2} m \mathbf{v}_s^2 = \frac{1}{2} \frac{m}{e^2 n_s} \mathbf{J}^2 = \frac{1}{2} \mu_0 \lambda^2 \mathbf{J}^2 \; . \tag{4.22}$$

This energy has to be smaller than the energy gained in the superconducting state. Thus the current has to be smaller than J_{cL}, where

$$J_{cL} = \frac{H_c}{\lambda} \; . \tag{4.23a}$$

By applying Ginzburg-Landau equations, we can obtain the same depairing current, except for a different numerical factor in front:

$$J_{cGL} = \frac{2\sqrt{2}}{3\sqrt{3}} \frac{H_c}{\lambda} \; . \tag{4.23b}$$

The Ginzburg-Landau and BCS approaches give the same relation for J_c with somewhat different numerical prefactors.

Thus the maximum current that can theoretically be sustained in a superconductor is of the order H_c/λ. As B_c is of the order 0.1 T in a conventional superconductor and λ is of the order 1000 Å, we get

$$J_c \sim 10^{12} \; \mathrm{Am^{-2}} = 10^8 \; \mathrm{Acm^{-2}} \; .$$

This is the maximum critical current that one can expect in a conventional type-II superconductor if we were able to completely solve the problem of the motion of vortices. Hence, the actual maximum current of $\sim 10^6\text{--}10^7 \; \mathrm{Acm^{-2}}$ in real materials is one or two orders of magnitude lower than the depairing current .

4.5.2. Different regimes

As we have already shown, an isolated vortex in a current $\mathbf{J}$ is subjected to a force per unit volume, often called the Lorentz force (see Figure 4.1b):

$$\mathbf{f} = \mathbf{J} \wedge \boldsymbol{\Phi}_0 \; . \tag{4.24}$$

Thus the lattice is subject to a force density per unit volume:

$$\begin{aligned} \mathbf{F} &= \mathbf{J} \wedge n\boldsymbol{\Phi}_0 \\ &= \mathbf{J} \wedge \mathbf{B} \; . \end{aligned} \tag{4.25}$$

This force $\mathbf{F}$ tends to set in motion the lattice of flux lines. If the vortices can move freely, it is not possible to pass a current above B_{c1} without energy dissipation. Therefore, to achieve a finite critical current, we have to pin the vortices, i.e., to find mechanisms or geometries which prevent flux lines motion. Let us call F_p the *average pinning force density* which prevents the lattice moving. If $F < F_p$, the lattice will not move and we have a non-dissipative current. If $F > F_p$, the lattice will move and we have the so-called flux flow regime. For $F = F_p$ we have the critical regime. If we define J_c by

$$F_p = J_c B \ , \qquad (4.26)$$

then J_c is the critical current of the material at zero temperature. For $J < J_c$ there is no dissipation, while for $J > J_c$ there is dissipation. The problem that we have to discuss is how to obtain and calculate this quantity F_p that we have just introduced under the term of average pinning force density.

In conventional superconductors the pinning of vortices usually occurs at the inhomogeneities in the material: due to the local variation of ξ or λ the energy of the flux tube changes accordingly and the vortex gets pinned to the energetically more favorable sites. In order to be efficient the inhomogeneities have to be of the order of ξ or λ, i.e. 100 to 1000 Å in conventional superconductors. The contrary seems to be true for new oxide superconductors where $\xi \sim 10$ Å. We shall therefore discuss the pinning problem in oxide superconductors separately in Chapter 7.

Let us now briefly describe the essential features of different regimes that we shall discuss in the remainder of this chapter:

FLUX FLOW: This is a dissipating regime where the flux line lattice moves and dissipation is nearly ohmic. It is the dominant regime in conventional superconductors when the current exceeds the critical current: $J > J_c$. The flux flow resistivity is linear in B.

HYSTERESIS: If $J < J_c$, the superconductor is in a metastable state. This results in the hysteresis of the magnetization. The amplitude of the hysteresis is directly related to the critical current as we will show in the Bean model (Section 4.7). Thus the critical current density can be estimated from magnetic measurements.

FLUX CREEP: Let us now consider the effect of temperature. In conventional superconductors, if the current is slightly lower than the critical current and the temperature is sufficiently high, thermal fluctuations permit the flux lines to move. This is the flux creep regime; the resistivity is not ohmic and depends strongly on temperature.

TAFF: Thermally assisted flux flow appears only in high temperature superconducting oxides for $J \ll J_c$. Although it is of the same nature as the flux creep, it has been given a new name (often abbreviated) in the context of high-T_c oxides: TAFF.

The aforementioned regimes are illustrated in Figure 4.4 which shows the current-voltage characteristics of type-II superconductors.

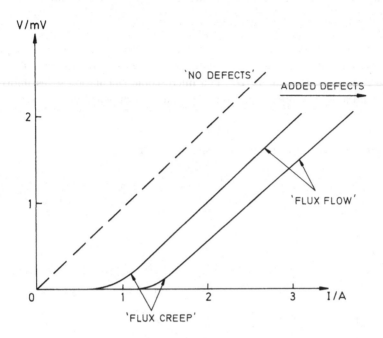

Figure 4.4: Schematic presentation of the current-voltage characteristics for two samples with different concentration of defects; critical currents are also different (after Strnad *et al.* 1964).

4.6. Flux Flow

We want to calculate the resistivity in the presence of a current larger than the critical current. We take the limit where the Lorentz force is large compared to the pinning force, so we can neglect the latter. The motion of a flux line dissipates energy so we can describe this dissipation in terms of viscosity. The vortex motion is assumed to be damped by a force proportional to the velocity. We can calculate the drift velocity of the vortex line as a function of this viscosity, η, by equating the Lorentz force to the friction force,

$$\mathbf{J}\Phi_0 = \eta \mathbf{v} \ . \tag{4.27}$$

The motion of the lattice induces an electric field $\mathbf{E}$ parallel to $\mathbf{J}$, given by the

Maxwell equation

$$\mathbf{E} = \mathbf{B} \wedge \mathbf{v} \ . \tag{4.28}$$

Using these two relations, one obtains the resistance of the flux flow

$$\rho_f = \frac{E}{J} = B \frac{\Phi_0}{\eta} \ . \tag{4.29}$$

The viscosity η is an unknown parameter. Experimentally it does not depend on the current and does not vary very much with B. Thus we obtain an ohmic regime and the resistivity is linear in B. The microscopic calculation gives information about η and a simple interpretation of this ohmic regime as it leads to

$$\rho_f = \rho_n \frac{B}{B_{c2}} \ . \tag{4.30}$$

Thus η is given in terms of the resistivity in the normal state:

$$\eta = \frac{\Phi_0 B_{c2}}{\rho_n} \ . \tag{4.31}$$

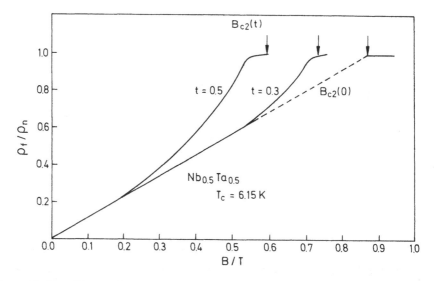

Figure 4.5: Flux flow resistivity as a function of B for different values of $t = T/T_c$ for a Nb-Ta sample. Dotted line represents the expected behavior for $t = 0$. Vertical pointers indicate value of $B_{c2}(t)$ (after Kim et al. 1964).

The resistivity in the flux flow regime is the same as that obtained for currents flowing inside the normal cores. Indeed, if we suppose that the vortex is a cylinder of normal metal with radius ξ then B/B_{c2} represents the fraction of the normal metal. In Figure 4.5 we show an example of flux flow resistivity as a function of field for different temperatures. The resistance increases as a function of the field and the temperature.

4.7. Hysteresis Cycle: The Bean Model

In Figure 4.2 we have already presented the magnetization curves as a function of the field. When there are some inhomogeneities that perturb the vortex motion, the magnetization curves are irreversible.

Cold drawing of metals creates metallurgical defects that are usually the pinning centers in conventional superconductors. The corresponding magnetization curve is strongly irreversible. One can carry out the sequence of annealing of the sample which reduces the number of defects and consequently the irreversibility of magnetization. Flux does not penetrate at B_{c1}, it remains pinned to the surface so the magnetization curve only gradually deviates from the straight line of perfect diamagnetism. On the contrary, the magnetization always disappears at B_{c2} independently of the irreversibility — the pinning forces always disappear at B_{c2}. Even when the field is lowered below B_{c2} the flux remains trapped within the specimen so it may exhibit paramagnetic ($B > \mu_0 H$), rather than diamagnetic, response.

The detailed mechanism of pinning is not clearly understood but all the structural features of the size of the vortex are effective pinning centers. One can create them in a controlled manner either by cold working of the metal, or by precipitating a second phase within the superconductor, or by irradiating the sample. In conventional superconductors the pinning increases with the number of inhomogeneities.

In order to interpret the irreversibility of the magnetization curves, Bean has developed a simple model that allows one to deduce the critical current of the sample. By using the Bean model one can estimate the critical current by measuring only the magnetization of the specimen.

We have seen that the force on the vortex is given by

$$\mathbf{f} = \mathbf{J}_{\text{ext}} \wedge \mathbf{\Phi}_0 . \tag{4.32}$$

Using Eq. (4.25) one can write the equation for the force per unit volume:

$$\mathbf{F} = \text{curl } \mathbf{H} \wedge \mathbf{B} . \tag{4.33}$$

To understand this equation we consider the one-dimensional case where H varies uniquely in one direction x:

$$F = \frac{dH}{dx} B . \tag{4.34}$$

Thus the force is related to the gradient of the field. It tends to make the field uniform in the sample.

Let F_p be the pinning force per unit volume. If $F > F_p$ equilibrium will not be achieved. The vortices will continue their motion until $F = F_p$ at all points. This physical regime is called *the critical state*. The variation of the field in the interior of the sample is directly linked to the current that the sample can sustain without vortex motion

$$F_p = J_c B = \frac{dH}{dx} B \; . \tag{4.35}$$

In its simplest form the Bean model supposes that the effect of vortex pinning is to determine the maximum gradient of the field. This gradient is equal to the critical current J_c.

4.7.1. Example of the slab

To illustrate this model let us consider the magnetization of a superconducting film of thickness d in the presence of a field parallel to its surface (Figure 4.6). Above B_{c1}, the field will begin to penetrate into the film up to a depth Δ determined by the critical condition

$$\frac{dH}{dx} = J_c$$

or

$$\Delta = \frac{H}{J_c} \; . \tag{4.36}$$

Above $B^* = \frac{1}{2}\mu_0 J_c d$, the field penetrates into the total depth of the film with a profile given in Figure 4.6a. B^* is the maximal external field which is completely screened in the middle of the superconductor. If the applied field is now reduced from $B_m \gg B^*$ one obtains the situation illustrated in Figure 4.6b. When the external field is brought to zero, a considerable quantity of flux still remains trapped in the interior of the sample. The flux within the sample will not disappear unless we apply an inverse external field $\frac{1}{2}B^*$.

Consequently one obtains hysteresis. If the external field is cycled with a maximum value $B_m < B^*$, one can calculate the energy dissipated during the cycle which is proportional to $(B_m)^3$. If the cycling is done with $B_m \gg B^*$ the dissipated energy is then proportional to B_m. These losses due to hysteresis limit the potential use of type-II superconductors in ac applications.

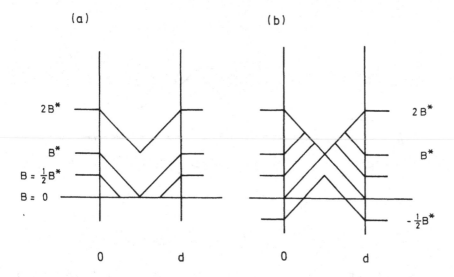

Figure 4.6: Profile of the field in the interior of a superconducting sample: a) for increasing field, b) for decreasing field.

The Bean model assumes that J_c is independent of B. Actually J_c has to be zero at B_{c2} where the pinning forces disappear. Another classical approximation is to take

$$J_c = \frac{\alpha}{B + B_0} \, , \tag{4.37}$$

where α and B_0 are adjustable parameters.

4.7.2. Application to measurements of critical current

It is easy to deduce $J_c(H)$ from the hysteresis cycles. The usual way is to take the difference, $\Delta M(H)$, between the magnetizations measured in increasing and decreasing fields ($M \uparrow - M \downarrow$) (see Figure 4.7). The variation of magnetization can be calculated in the Bean model (Figure 4.8). It is proportional to the area shown in this figure. Thus, we have

$$J_c(\text{Am}^{-2}) = 2\frac{\Delta M}{d} \quad \text{in MKSA} \, . \tag{4.38}$$

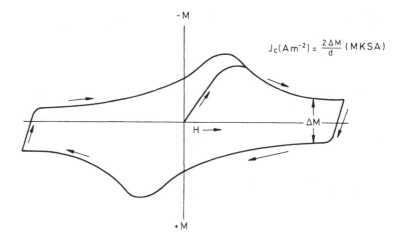

Figure 4.7: Typical magnetization curve for type-II superconductors. The hysteresis ΔM in a given field H is a measure of the critical current density J_c. d in the above relation denotes the sample thickness, or the grain size for the case of polycrystalline ceramic specimens (after Doss 1990).

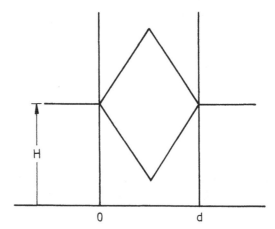

Figure 4.8: Modification of the profile of the field for a given external field for increasing and decreasing fields. The variation of the magnetization of the film is proportional to the area of the parallelepiped.

4.8. Pinning of Vortices

We have introduced the pinning force per unit volume F_p which is related to the measured critical current by Eq. (4.26):

$$F_p = J_c B \ .$$

It does not say anything about the mechanism of pinning nor how to increase the value of the critical current. Even J_c is not related to the pinning force of an individual vortex f_p. Thus if we want to increase F_p we have first to make the link with the pinning force on an individual vortex f_p and then study the mechanism which can pin an individual vortex line.

We assume random inhomogeneities in the bulk of the sample, which we call pinning points. These objects will be described by an interaction potential with a vortex line. The pinning points are inhomogeneities which can either favor or inhibit the pair condensation responsible for superconductivity, i.e., which repel or attract flux lines. The first problem that we will deal with is the way these point interactions between flux lines and pinning points add to determine the pinning force density F_p. One is tempted to write

$$F_p = N f_p \ , \tag{4.39}$$

where N is the number of interactions between pinning points and vortices. This is not correct in general. Indeed, consider that we have a rigid lattice of vortices and random inhomogeneities. We would have no pinning at all. The reason stems from the fact that the pinning forces are randomly oriented and statistically cancelled. One can also understand this result by considering the fact that the interaction energy in an infinite medium would be independent of the relative position of the rigid lattice of flux lines and the random array of pinning centers. As pinning effectively occurs an explication is needed. It is essential to pinning processes that the lattice be deformed. In this case the total energy of the system is lowered by deformation of the lattice and pinning may occur if an energy increase is required to move the lattice with respect to the pinning array. We mention also the opposite limit, i.e., if we have no lattice but a liquid of vortices. In that case we have also $J_c = 0$ because we would need to pin each vortex in order to prevent it from moving. Thus the lattice stiffness is central to the pinning problem. It is necessary to be able to describe the rigidity of the lattice.

In order to understand the effects of the pinning centers, we consider an ideal flux line lattice and a random distribution of pinning centers whose pin strength gradually increases from zero. For very small pinning force, the lattice will respond elastically, and by using the theory of elasticity of the flux line lattice, we could calculate the displacement of each vortex and the energy of the lattice. If we increase the pinning force, we shall lose the positional long range order of the lattice. Stronger pinning will create dislocations and other defects in the lattice. It is generally assumed that in all cases of interest, the flux line lattice can be split into elastically independent, correlated regions of volume V_c. The critical current is then obtained for a Lorentz force which can set up in motion this correlated region. In this region, the pinning force is random and the net effect is given by the fluctuations in the region. If we have n_v pinning centers per unit volume, this force will be given by

$$[n_v V_c \langle f_p^2 \rangle]^{\frac{1}{2}} . \tag{4.40}$$

Thus the pinning force density is given by this force divided by the volume V_c

$$F_p = \left[\frac{n_v}{V_c} \langle f_p^2 \rangle \right]^{\frac{1}{2}} . \tag{4.41}$$

The first interesting consequence of this expression is that only the square of the pinning force is involved. This is an important result: repulsive forces can also pin an array of vortices, not only the attractive ones.

However, the result depends not only on calculation of the microscopic pinning force f_p but also on the volume V_c of the correlated region. The calculation of V_c turns out to be very complicated as it depends on the elastic constants of the flux line lattice as well as on the strength of the pinning center. It is important to note that the pinning force increases with decreasing volume V_c.

4.8.1. Elasticity of the flux line lattice

For a complete description of the elastic properties, we have only three moduli, C_{11}, C_{66} and C_{44} in Voigt's notation of elasticity theory. The compressional modulus C_{11} represents the deformation that changes only the size of the lattice parameter and not its shape. It is large compared with the shear modulus C_{66} and the tilt modulus C_{44}, and it is therefore often neglected in the analysis of deformation of the lattice. C_{44}, the tilt modulus, describes deformation that tilts a bundle of flux lines array from the field direction while leaving its cross section in the xy-plane constant. C_{66} is the shear modulus in a plane perpendicular to the field.

Short-scale distortion of the flux line lattice cannot be described by elasticity theory, since the magnetic field does not change over distances smaller than the penetration depth. As a result, there is a wavenumber dependence of the elastic moduli.

We shall not give any derivation but just quote some important results. i) The compressional modulus and the tilt modulus are finite and they are even increasing up to B_{c2}. ii) On the contrary the shear modulus vanishes at B_{c2}. This indicates a tendency of the lattice to develop fluid-like behavior near B_{c2}, instead of forming a regular lattice.

4.8.2. Correlated volume

The first important parameter in the calculation of the pinning force density F_p is the estimate of the volume within which the vortex lattice is almost regular. The lattice disorder is different along the z-axis (the field direction) and in the xy-plane. Let us call L_c the length along the flux line and R_c the length in the plane. We have

$$V_c = L_c R_c^2 \ . \tag{4.42}$$

The ratio L_c/R_c is in most cases a function of the ratio of the tilt and shear moduli

$$\frac{L_c}{R_c} \propto \sqrt{\frac{C_{44}}{C_{66}}} \ . \tag{4.43}$$

In conventional superconductors the tilt modulus is larger than the shear modulus, $L_c \gg R_c$. Disorder is much more difficult along the field direction than perpendicular to it. In thin films with the field perpendicular to the film, L_c is often larger than the thickness of the film and the flux line lattice remains ordered along the field direction, the flux lines are straight lines.

In strongly disordered case, R_c is of the order of the lattice spacing as we have a nearly amorphous lattice. The simplest approximation for L_c is to make it equal to the coherence length. However the validity of this approximation is not known and a number of approximations involving the elastic constant and the pinning strength, can be found in the literature.

4.8.3. Elementary pinning force

The origin of the inhomogeneities in the order parameter leading to pinning is a very complex subject, involving point defects and such large scale defects as grain boundaries, twin boundaries, dislocations, precipitates, etc. ... . For technical superconductors, the important classes of defects are grain boundaries as in A-15 and Chevrel phase compounds or in Nb-N, and precipitates as in Nb-Ti.

Crystal defects give rise to local variations of the superconducting order parameter. These local changes couple to the flux lines lattice. In a first classification of the elementary interactions, one can distinguish between magnetic and core interactions.

In the first class there are effects of surfaces parallel to the applied magnetic field. For instance, the current distribution around a vortex core near the surface is forced to change in order that the normal component of the superfluid current vanishes at the boundary. Theoretically, this is achieved by assuming an anti-vortex image at the other side of the boundary. This results in an attractive surface vortex interaction. The net effect is a potential barrier for flux entry or exit. Another example of this magnetic interaction is the thickness variation of thin film for a field perpendicular to the surface. The vortices are pinned at points of smallest thickness, where the line energy is minimum. The typical length for this class is the penetration depth λ. In materials with a large Ginzburg-Landau parameter this kind of interaction is small and disappears with increasing magnetic field.

In the second class, the coupling to the change of the order parameter is the origin of flux pinning for most defects. Defects are regions which deviate from the surrounding material for certain properties: for instance, density, elasticity, electron-phonon coupling, mean free path. The first three properties lead to local changes of T_c, the last to a local change of κ. Thus they are often referred to as δT_c or $\delta \kappa$ pinning. The typical length scale is now the coherence length.

Many calculations of this elementary pinning force exist but nothing is as simple and definitive as can be discussed at our entry level in all these particular cases. In most cases the pinning force is proportional to Δ^2 as the condensation energy is given as a function of the square of the order parameter. Close to B_{c2} we have

$$\Delta^2 \sim 1 - \frac{B}{B_{c2}} \ . \tag{4.44}$$

The pinning force always vanishes at B_{c2}, and in most cases it vanishes linearly with the field. In that case, it is a fraction of the condensation energy divided by a typical length.

4.9. Flux Creep

Flux line motion is due to the Lorentz force density which acts perpendicularly to the flux lines. The flux lines are pinned by defects which lead to a distortion of the flux line lattice even for no external current. At zero temperature, flux line motion is only possible if the Lorentz force density exceeds the average pinning force density. The current density J_{ex} is non-dissipative if $J_{ex} < J_c(0, B)$.

For $J_{ex} > J_c$, flux line motion with velocity v leads to an electric field and hence to a finite voltage V. This is the flux flow regime that we have already described.

However at a finite temperature there is a finite probability that the flux lines overcome the pinning energy barriers. Thus, even for $J_{ex} < J_c(T, B)$ we can have some motion and resistivity. This is called thermally activated flux creep. In the case when $J_{ex} \gg J_c(T, B)$ we have the flux flow already described.

Flux creep depends on various parameters that are not well known, but we can make a qualitative analysis of this phenomenon in a simple way. In fact in most cases one observes the motion of bundles of flux lines. This is a thermally activated jump from one location to another one nearby. In the simplest case the activation (or barrier) free energy $U_0(T, B)$ which is overcome by the jump of a flux line bundle is determined by the condensation energy density gained by the flux lines in a favorable region multiplied by a suitable volume.

At zero temperature, the barrier energy $U_0(0, B)$ is connected to the critical current density $J_c(0, B)$. Since for $J_{ext} = J_c$ the Lorentz force density has to be equal to the pinning force density, we have

$$U_0(0, B) = J_c(0, B)BV_c d , \qquad (4.45)$$

where d is the distance over which the flux bundle moves; it is of the order a_o.

In conventional superconductors, one always has $U_0(T, B) \gg k_B T$ for all temperatures $T \leq T_c$. U_0/k_B ranges approximately between 100 and 1000 K: flux creep is a very small phenomenon.

(a)   (b)

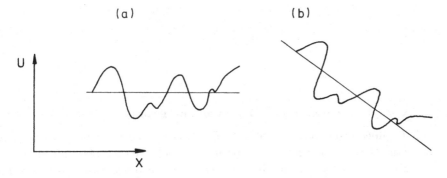

Figure 4.9: Schematic presentation of the energy of the bundle of flux lines as a function of position: a) without, and b) with current.

In Figure 4.9 we show the energy of the bundle of flux lines as a function of position. Due to thermal fluctuations there is a possibility for a bundle to jump over the barrier to an adjacent pinning site, i.e., an adjacent potential well. This possibility is given by the Arrhenius law. If there is an applied external current J_{ex}, the flux bundle will move with a velocity given by

$$v = 2v_0 \exp-\frac{U_0}{k_BT} \sinh\frac{BJ_{\text{ex}}V_cd}{k_BT}$$

$$\sim v_0 \exp-\frac{1}{k_BT}(U_0 - BJ_{\text{ex}}V_cd) \ . \tag{4.46}$$

In this formula v_0 is a microscopic velocity proportional to the attempt frequency to jump over the barrier. It is not accurately known but may be in the range of 10 m/s in conventional superconductors. The flux creep velocity v generates an electric field $E = Bv$, hence one can observe resistivity. Now if E_c is the smallest electric field that can be measured, say 1μV/cm, as long as the flux creep velocity v does not give rise to this field, one does not observe dissipation. The critical current $J_c(T)$ is obtained for E_c by

$$E_c = Bv = Bv_0 \exp-\frac{1}{k_BT}(U_0 - BJ_c(T)V_cd) \ , \tag{4.47}$$

or

$$J_c(T) = J_c(0) \left(1 - \frac{k_BT}{U_0} \ln\frac{Bv_0}{E_c}\right) \tag{4.48}$$

where $J_c(0)$ is given by Eq. (4.45).

In conventional superconductors, flux creep occurs only when the current is large enough to nearly overcome the barrier, i.e., when

$$\frac{J_c - J}{J_c} \ll 1 \ . \tag{4.49}$$

4.9.1. New phenomena in high-T_c superconductors

In high-T_c superconductors, the ratio $U_0(B,T)/k_BT$ is considerably smaller than in conventional superconductors. This occurs first because U_0 is determined by the condensation energy times a suitable volume which obviously must depend on the coherence length. This length is much smaller in high-T_c superconductors, thus U_0 is smaller. Second the temperature is higher than in conventional superconductors. Thus flux creep cannot be a small effect and sometimes one speaks about giant flux creep phenomena.

Flux creep, as we described it, is a phenomenon that occurs when the driving force is almost equal to the pinning force, i.e. just below the flux flow regime. But if the pinning barrier is small and the temperature high enough to overcome the barrier, this effect can be observed in the limit of small driving forces. In order to distinguish the conventional flux creep (which occurs under the condition given by Eq. (4.49)) from this effect, which occurs for $J \ll J_c$, a new name has been found: the *thermally activated flux flow* (TAFF). TAFF is observable only when the potential barrier is low and gives rise to a resistivity (Figure 4.10):

$$\rho \sim \rho_0 \exp{-\frac{U_0}{k_B T}} \ . \tag{4.50}$$

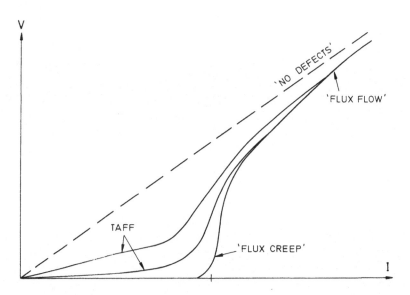

Figure 4.10. Schematic illustration of voltage-current characteristic for different regimes.

So far we have discussed the Bean model and shown how to deduce the critical current from the magnetization curve. However, if there is flux creep, the magnetization will relax towards the equilibrium magnetization. It can be shown that for long times, the relaxation law is logarithmic:

$$M(t) \sim -\ln{\frac{t}{t_0}} \quad \text{for} \ \ t > t_0 \ . \tag{4.51}$$

Hence the measurements of relaxation allow one to estimate the barrier U_0. If the free energy barrier is small, relaxation can occur during the time of measurements; this occurs in some high-T_c oxides. Above a given temperature, the magnetization relaxes to its equilibrium value and one obtains a reversible magnetization curve. Thus one would deduce the zero critical current by using the Bean method, Eq. (4.38).

However, if we pass the transport current and directly measure the voltage, we can detect a signal if the current is higher than some critical value which corresponds to E_c. So, by carrying out the transport measurements, we would measure finite critical current although an analysis of magnetization measurements based on Eq. (4.38) would make it appear zero.

Summary

1. Between B_{c1} and B_{c2}, flux enters into type-II superconductor and forms a triangular lattice of flux lines, the **vortex lattice**.

2. The Lorentz force $\mathbf{F}$ on a given flux line, due to a current $\mathbf{J}$, is given by

$$\mathbf{F} = \mathbf{J} \wedge \mathbf{\Phi}_0 \; .$$

3. The depairing current of a superconductor is given as

$$J_c \text{ (depairing)} \approx \frac{H_c}{\lambda} \; .$$

J_c is $\sim 10^8$ and 10^9 Acm^{-2} for conventional and oxide superconductors respectively.

4. The critical current is related to the average pinning force density F_p by

$$F_p = J_c B \; .$$

5. If the flux line lattice moves uniformly in the presence of a current, this regime is called the *flux flow* and the resistivity is

$$\rho = \rho_n \frac{B}{B_{c2}} \; .$$

6. The hysteresis cycle enables one to estimate the critical current via the Bean formula (in MKSA units)

$$J_c = 2\frac{\Delta M}{d} \; .$$

7. Flux creep is the thermal hopping over an energy barrier of a bundle of flux lines within a correlated volume. This effect is important when the energy barriers are comparable to the thermal energy $k_B T$ as in the case of high-T_c oxides.

Further Reading

A. M. Campbell and J. E. Evetts: *Critical Currents in Superconductors*, Taylor and Francis, London, 1972

S. Foner and B. B. Swartz: *Superconductor Materials Science, Metallurgy, Fabrication and Applications*, Plenum, New York, 1981

P. G. de Gennes: *Superconductivity of Metals And Alloys*, W.A. Benjamin, New York, 1966

T. Luhman and D. Dew-Hughes (editors): *Metallurgy of Superconducting Materials*, Academic Press, New York, 1979

D. Saint-James, G. Sarma and E. J. Thomas: *Type-II Superconductivity*, Pergamon Press, Oxford, 1969

D. R. Tilley and J. Tilley: *Superfluidity and Superconductivity*, Adam-Hilger, 1990

M. Tinkham: *Introduction To Superconductivity*, McGraw-Hill, New York, 1975; reprinted by Robert E. Krieger, Malabar, Florida, 1985

M. N. Wilson: *Superconducting Magnets*, Oxford, 1983

Chapter 5. RESULTS FROM THE MICROSCOPIC THEORY

Preview

The Bardeen, Cooper and Schrieffer (BCS) theory relies on an attractive interaction between electrons. In order to explain how phonons can lead to such an unusual interaction we first introduce the concept of dielectric constant. We then sketch the BCS theory focussing essentially on the main results. The microscopic BCS theory permits one to wholly justify the phenomenological Ginzburg-Landau approach discussed in Chapter 3.

5.1. Introduction

The formulation of a theory of superconductivity is very difficult because of the smallness of the energy involved in the process. This energy, as calculated from the condensation energy, can be as small as 10^{-8} eV per atom. For comparison, the cohesive energy of a metal is of the order of several eV per atom. The difference in energy between two crystallographic structures, fcc and hcp, for instance, is of the order of 10^{-2} eV. Thus, one has to find what interaction can lead to a state with such a small difference in energy (as compared with the normal state) and to such strikingly different properties.

The first clue to understanding the mechanism of conventional superconductivity arose from the discovery of the isotope effect: two different isotopes of the same metal exhibit different T_c's. The relation, valid for some simple metals, is given by:

$$T_c \sim M^{-\alpha} ,$$

where M is the atomic mass of the isotope and α is roughly 0.5. Why is the mass of an atom involved in a purely electronic property? Obviously, the motion of ions has something to do with superconductivity.

The second clue was found by Leon Cooper. He showed that a normal metal (with 'standard' metallic properties) could not be formed if there was a small attraction between electrons. In such a case, two electrons would form pairs however small the attractive interaction. And, if electrons did form pairs, completely different properties for the whole ensemble of electrons would be observed.

Thus, the central idea is that motion of ions can lead to attractive interaction between electrons. How can this happen? When an electron moves among the positive ions of the lattice, it attracts them. However, the motion of ions is slow so the electron advances a great deal while the ions somewhat converge towards each other. They build a region of positive charge, which before relaxing attracts another electron. So, due to this slow response of the ions, there appears an effective attractive interaction between electrons. Of course the repulsive Coulomb interaction is also present and there is a delicate 'competition' between the two interactions which makes some metals superconducting and others non-superconducting.

In this chapter, we first try to understand a little better this attractive mechanism by studying the response of a normal metal to a small perturbation, i.e., a test charge introduced into the metal: this leads to the concept of a dielectric 'constant'. We show how a change of sign of the dielectric 'constant' changes a repulsive interaction into an attractive one and consequently leads to pairing. In the second part of the chapter, we sum up and briefly discuss the main results of the microscopic BCS theory, which we cannot discuss in detail at the level of this textbook.

5.2. Normal Metal

In a normal metal, current can flow easily. This means that electrons are nearly free to propagate. One of the simplest models of the metal is a box filled with a gas of free electrons. This picture is due to Sommerfeld and works remarkably well (considering that it is an oversimplification). In quantum mechanics, an electron of energy E and the momentum $\mathbf{p}$ is described by a wavefunction

$$\Psi(\mathbf{r}) = e^{i\mathbf{k}\mathbf{r} - i\omega t} , \qquad (5.1)$$

where $\mathbf{k}$ and ω are given by the de Broglie formulae

$$\omega = \frac{E}{\hbar}$$

and

$$|\mathbf{k}| = \frac{|\mathbf{p}|}{\hbar} = \frac{2\pi}{\lambda} . \qquad (5.2)$$

Within the box, the waves have to be stationary, which means that boundary conditions are to be added. A convenient way is to introduce the Born-Von Karman boundary conditions. If L_x is the length of the box along the x-axis, possible values of k_x are given by

$$k_x = \frac{2\pi}{L_x} n_x ,$$

where n_x is an integer. Thus such a state is defined by three integers: n_x, n_y, n_z. Electrons are fermions. Therefore, for a given wave vector one can have only two electrons in a given state: one with spin up and the other with spin down. To accommodate all the electrons in the box, we have to fill all the states with wave vectors $\mathbf{k}$:

$$|\mathbf{k}| < k_{\mathrm{F}} ,$$

where k_F is the Fermi wave vector which is related to the density of electrons, n, by

$$n = \frac{k_F^3}{3\pi^2} \ . \tag{5.3}$$

The dispersion relation between the energy and the wave vector is given by the simple (free electron) approximation

$$E = \frac{p^2}{2m} = \frac{\hbar^2 k^2}{2m} \ . \tag{5.4}$$

Thus all the states below the Fermi energy E_F are filled. At this energy the density of states is given by

$$N(E_F) = \frac{m k_F}{2\pi^2 \hbar^2} \ . \tag{5.5}$$

The actual density is of the order of 10^{23} electrons per cm^3 so the Fermi energy is of the order of a few eV. Some typical values are given in Table 5.1.

Table 5.1: Some typical values of the characteristic quantities of a few normal metals and a high-T_c oxide, YBa$_2$Cu$_3$O$_{6.9}$ (YBCO).

	$v_F/10^8$cm/s	$k_F/10^8$cm^{-1}	$n/10^{22}$/cm^3	E_F/eV	Θ_D/K
Cu	1.57	1.36	8.47	7.00	340
Al	2.03	1.75	18.10	11.70	390
Nb	1.37	1.18	5.56	5.32	320
YBa$_2$Cu$_3$O$_{6.9}$	0.1	$\sim$0.5*	0.7	0.1	$\sim$400

*an estimate within free electron model.

That electrons behave very much like free electrons found confirmation in measurements of susceptibility or specific heat. The susceptibility is constant at low temperatures and is given by

$$\chi = 2\mu_B^2\, N(E_F) \ . \tag{5.6}$$

The electronic specific heat is linear in temperature:

$$C_v = \gamma_T \ ,$$

where

$$\gamma = \frac{2\pi^2}{3}\, N(E_F)\, k_B^2 \ . \tag{5.7}$$

Such behavior is indeed observed in most metals, which is rather surprising as this model is oversimplified.

Hence, one may ask two questions: first, what happened to the lattice of ions? In fact, qualitatively the Bloch theorem[†] asserts that the lattice does not change things very much: the energy becomes a periodic function of k. For instance, if the ions form a simple cubic lattice with an interatomic distance a the dispersion or E–k relation becomes

$$E = zt \left(\cos k_x a + \cos k_y a + \cos k_z a \right) , \tag{5.8}$$

which for small k has the same k^2 dependence as in Eq. (5.4). One can improve our simple model by replacing, in Eq. (5.4), the electron mass m with an effective electron mass m^*.

The second fundamental question concerns the fact that we have completely neglected the Coulomb repulsion between electrons. How can one justify the fact that despite this neglect we obtain sensible results? The answer is that the average kinetic energy of the electrons is large as compared with the repulsive potential energy. The average kinetic energy of the gas is of the order of E_F, i.e., few eV and it is proportional to $n^{2/3}$. The potential energy is of the order of $\frac{e^2}{r_e}$, where r_e is the average distance between two electrons and so is proportional to $n^{1/3}$. For large values of n, the repulsive term is negligible compared with the kinetic one. This is the situation in most metals, but this term becomes important and even predominant in the low density limit. In most conventional superconductors this term is not predominant and one can describe them with our simple approximation.

5.3. Instability of the Normal State

It was Leon Cooper who noticed that the gas of electrons, described above, if stable in the presence of repulsion between electrons, is completely unstable in the case of attraction between them. In his analysis, Cooper adds two electrons of opposite wave vectors just above the Fermi surface and looks for a wavefunction for those two electrons of the form

$$\Psi(\mathbf{r}_1 - \mathbf{r}_2) = \sum_{\mathbf{k}} g(\mathbf{k}) e^{i\mathbf{k}(\mathbf{r}_1 - \mathbf{r}_2)} , \tag{5.9}$$

[†] The Bloch theorem states that the wavefunction of an electron moving in a periodic potential of the lattice has the form of a plane wave, $e^{i\mathbf{k}\mathbf{r}}$, modulated by a function $u(\mathbf{k}, \mathbf{r})$ that has the same periodicity as the periodic potential

$$\Psi(\mathbf{k}, \mathbf{r}) = e^{i\mathbf{k}\mathbf{r}} u(\mathbf{k}, \mathbf{r}) .$$

where $g(\mathbf{k})$ is the probability amplitude for finding one electron with momentum $\hbar\mathbf{k}$ and the corresponding electron with momentum $-\hbar\mathbf{k}$.

Of course,

$$g(\mathbf{k}) = 0 \quad \text{for} \quad |\mathbf{k}| < k_F \ , \tag{5.10}$$

as all the states below k_F are already filled with electrons.

Cooper assumed an attraction between these two electrons and solved the Schrödinger equation. His objective was to calculate the wavefunction and the energy of the pair which can be measured from the Fermi energy

$$E = 2E_F + \varepsilon \ . \tag{5.11}$$

He discovered that the *pair* always *binds*, i.e. the energy ε is negative and is given by the relation

$$\varepsilon = -2\hbar\omega_D \exp\left[-\frac{2}{N(E_F)V} \right] , \tag{5.12}$$

where $\hbar\omega_D$ is a characteristic energy of the attractive potential and V is the strength of this attractive potential.

This is not an obvious result, for if you have two particles which attract, they will not bind in general, unless the attractive potential is large enough. Here they *always bind even if V is very small*! These two electrons constitute what is now known as a Cooper pair. The binding energy is large only if $N(E_F)V$ is large (as it enters into the exponential). So one needs either large V or large density of states at the Fermi level, $N(E_F)$. If one calculates the extension ξ of the pair, i.e., the average distance between two electrons, one finds

$$\xi \sim \frac{\hbar v_F}{|\varepsilon|} \ .$$

As ε can be small this quantity can be large, i.e., hundreds of angstroms.

What are the consequences of all electrons near the Fermi energy attracting each other (and not only the two added ones as in Cooper's analysis)? They will tend to form pairs in order to decrease their energy. Thus, our model of filling the k-states up to the Fermi level with electrons breaks down completely. In the presence of a weak attraction between electrons the ground state which we used for describing the normal metal is no longer the ground state of the system so we have to think of some other ground states.

5.4. Concept of Dielectric 'Constant'

In order to understand the origin of this unusual attraction between electrons we will proceed as follows. We shall first describe the reaction of the charged medium to a test charge in terms of the dielectric 'constant'. In a simple metal electrons screen a test charge over a length called Thomas-Fermi length. We shall see in Section 5.5 that if we consider a gas of electrons in the Thomas-Fermi approximation the dielectric constant cannot turn the repulsive potential into an attractive one. However, if we also add the lattice of ions and bring in phonons or elastic waves to the medium, the reaction is different (Section 5.6): the dielectric constant can become *negative*. In that case the sign of interaction can be turned into that of an *attractive* one.

In an electron gas, if one introduces a test charge at point **r** the electrons will rearrange themselves in order to screen the field of this added charge. The response of the system is described by Maxwell equations with the introduction of the dielectric constant. Indeed, the displacement vector **D** is related to the external applied charge density, our test charge ρ_t, by

$$\text{div } \mathbf{D} = \rho_t \ . \tag{5.13}$$

The electric field vector is related to the total charge density

$$\text{div } \mathbf{E} = \frac{\rho_{\text{tot}}}{\varepsilon_0} = \frac{\rho_t + \rho}{\varepsilon_0} \ , \tag{5.14}$$

where ρ is the induced charge in the system due to the test charge ρ_t. The dielectric constant is usually defined as

$$\mathbf{D} = \varepsilon_0 \varepsilon \, \mathbf{E} \ , \tag{5.15}$$

where ε_o is the permittivity of free space. Hence we get for the dielectric constant

$$\varepsilon = \frac{\rho_t}{\rho_t + \rho} \ . \tag{5.16}$$

If the test charge is varying in space and time, one defines the dielectric constant, $\varepsilon(\mathbf{q}, \omega)$ which depends on the wavelength and frequency. From Poisson's equation one can easily show that

$$\varepsilon(\mathbf{q}, \omega) = \frac{V_{\text{ext}} (\mathbf{q}, \omega)}{V(\mathbf{q}, \omega)} \ , \tag{5.17}$$

where $V_{\text{ext}}(\mathbf{q}, \omega)$ is the Fourier component of the electrostatic potential of the test charge in vacuum, and $V(\mathbf{q}, \omega)$ is the Fourier component of the potential of the test charge in the medium.

5.5. Dielectric Constant of a Gas of Electrons

We expect that the electrical potential of the test charge in an electron gas is screened exponentially over the screening length l_s:

$$V(r) = \frac{q}{4\pi \varepsilon_e \varepsilon_0 r} \, e^{-r/l_s} \ . \tag{5.18}$$

It means that the dielectric constant of the electron gas is given by

$$\varepsilon_e(q) = 1 + \frac{k_{\mathrm{TF}}^2}{q^2} \ . \tag{5.19}$$

Thomas–Fermi theory gives for k_{TF}

$$k_{\mathrm{TF}}^2 = l_s^{-2} = \frac{3}{2} \frac{ne^2}{E_{\mathrm{F}}} \frac{1}{\varepsilon_o} = \frac{4}{\pi} \frac{k_{\mathrm{F}}}{a_o} \ , \tag{5.20}$$

where a_o, the Bohr radius, is ~ 0.5 Å. The magnitude of the screening length l_s is of the order of the Bohr radius.

This is the simplest approximation for the dielectric constant of a gas and it is called the Thomas–Fermi approximation. It is obvious that $\varepsilon_e(\mathbf{q})$ is *always positive*, thus the screened potential is always of the same sign as the bare potential $V_{\mathrm{ext}}(\mathbf{q}, \omega)$ of the test charge. This dielectric constant cannot turn a repulsive potential into an attractive one. Let us see what happens in a more realistic case where we add the positive ions of the lattice.

5.6. Motion of Ions in Metal

Consider now a lattice of positive ions embedded in a gas of electrons. The ions are not rigid but are vibrating around their equilibrium positions. In fact there are elastic waves propagating in the solid at all temperatures. As this motion is responsible for the conventional superconductivity, we will now study these waves.

Let us consider ions of charge Ze. If n is the electron density in the solid, the average ion density is n/Z. If M is the mass of an ion, one can write its equation of motion as

$$M \frac{d\mathbf{v}_i}{dt} = Ze\mathbf{E} \ . \tag{5.21}$$

In analogy with the electron current we introduce

$$\mathbf{J}_i = \frac{n}{Z} \cdot Ze \cdot \mathbf{v}_i = ne\,\mathbf{v}_i \ . \tag{5.22}$$

Thus we have

$$\frac{d\mathbf{J}_i}{dt} = nZe^2\,\mathbf{E} \ . \tag{5.23}$$

If ρ_i is the density of ions, the equation of conservation for the number of ions reads

$$\frac{\delta \rho_i}{\delta t} + \text{div } \mathbf{J}_i = 0 \ . \tag{5.24}$$

From the last two equations we get

$$\frac{\delta^2 \rho_i}{\delta t^2} = -\frac{nZe^2}{M} \text{ div } \mathbf{E} \ , \tag{5.25}$$

so Maxwell equations (see Appendix A) give

$$\text{div } \mathbf{E} = \frac{\rho_i + \rho_e + \rho_t}{\varepsilon_o} \ ,$$

where ρ_e is the density of electrons. Finally we get

$$\frac{\delta^2 \rho_i}{\delta t^2} = -\Omega_i^2 \left(\rho_i + \rho_e + \rho_t \right) \ , \tag{5.26}$$

where

$$\Omega_i^2 = \frac{nZe^2}{M\varepsilon_o} \ , \tag{5.27}$$

is the ionic *plasma* frequency.
The dielectric constant of electrons is now given by

$$\varepsilon_e = \frac{\rho_i + \rho_t}{\rho_e + \rho_i + \rho_t} \ , \tag{5.28}$$

as $\rho_i + \rho_t$ is now the test charge for electrons. Substituting Eq. (5.28) in (5.26) we obtain an important equation:

$$\frac{\delta^2 \rho_i}{\delta t^2} = -\frac{\Omega_i^2}{\varepsilon_e(q)} \left(\rho_i + \rho_t \right) \ . \tag{5.29}$$

This equation tells us that without a test charge the ions are moving like waves. These waves, called phonons, have a wavelength dependent frequency. The ions move with a frequency ω_q given by

$$\omega_q^2 = \frac{\Omega_i^2}{\varepsilon_e(q)} \ . \tag{5.30}$$

For small q we get

$$\omega_q = \frac{\Omega_i}{\sqrt{\varepsilon_e(q)}} \sim \frac{\Omega_i}{k_{\text{TF}}} q \ . \tag{5.31}$$

This gives the velocity of sound in a metal. By substituting for Ω_i and k_{TF} we find

$$v_s = v_F \sqrt{\frac{Z}{3} \frac{m}{M}} . \tag{5.32}$$

This means that the velocity of sound in a metal is of the order of the Fermi velocity divided by ~ 100.

5.6.1. Typical phonon frequency: Debye frequency

What is the order of magnitude of the typical phonon frequency and typical phonon wavelength? If we assume that the sound velocity is frequency-independent (which is called the Debye approximation), i.e.,

$$\omega_q = v_s q , \tag{5.33}$$

we can calculate the maximum wave vector q_D and the maximum frequency ω_D (the Debye wave vector and Debye frequency). Indeed, the total number of modes for phonons is $3N$ if N is the number of atoms, i.e., N modes for each given polarization. The total number of modes whose wave vector is smaller than q_D is given by the volume of a sphere of radius q_D divided by $\frac{(2\pi)^3}{V}$:

$$N = \frac{V}{(2\pi)^3} \frac{4\pi}{3} q_D^3 . \tag{5.34}$$

By setting $V_0 = \frac{V}{N}$ for the atomic volume, the Debye wave vector is

$$q_D^3 = \frac{1}{V_0} 6\pi^2 \tag{5.35}$$

and the corresponding Debye frequency is

$$\omega_D^3 = \frac{1}{V_0} 6\pi^2 v_s^3 . \tag{5.36}$$

These two formulas give the order of magnitude of the frequency and wave vector. Typically ω_D is of the order of 10^{12} to 10^{13} s^{-1}. We can define the Debye temperature by the relation

$$\hbar\omega_D = k_B \Theta_D . \tag{5.37}$$

One can easily estimate that the corresponding Debye temperature is of order of 10^2 K (see also Tables 5.1 and 5.2).

Legend:

Symbol	Quantity	Units
Critical temperature, T_c		Kelvin
Debye temperature, θ_D		Kelvin
Electron specific heat		mJ/mole K
Electron-phonon coupling		dimensionless
Density of states $N(E_F)$		states/atom eV

1	2	3	4	5	6	7	8	9	10	11	12	3	4	5	6	7	8
Li FILM	Be 0.03																
												Al 1.2 423 1.4	Si FILM PRES	P PRES			
		Sc 0.01 470 10.9	Ti 0.4 415 3.3 0.54 1.4	V 5.4 383 9.8 1.0 2.1	Cr FILM						Zn 0.9 316 0.7	Ga 1.1 317 0.60	Ge FILM PRES	As PRES	Se PRES		
		Y PRES	Zr 0.6 290 2.8 0.22 0.8	Nb 9.3 276 7.8 0.85 2.0	Mo 0.9 460 1.8 0.35 06	Tc 7.8 411 6.3	Ru 0.5 580 2.8 0.47 0.9		Pd IRRAD		Cd 0.5 210 0.67	In 3.4 108 1.7	Sn 3.7 196 1.8 (W)	Sb PRES	Te PRES		
Cs FILM PRES	Ba PRES	La 4.9 6.3 (α) (β)	Hf 0.1 252 2.2 0.14 0.8	Ta 4.4 258 6.2 0.75 1.7	W 0.02 383 0.9 0.25 0.5	Re 1.7 415 2.4 0.37 0.74	Os 0.7 500 2.4 0.44 0.68	Ir 0.1 425 3.2 0.4 0.35			Hg 4.2 75 1.8	Tl 2.4 88 1.5 0.8	Pb 7.2 102 3.1 1.55	Bi FILM PRES			

PRES : Pressure
IRRAD: Irradiated

Table 5.2: Characteristic values of superconducting elements (after Poole *et al.* 1988).

5.7. Origin of the Attractive Interaction

We shall now consider how a test charge is screened when embedded in a medium consisting of electrons and ions. The dielectric constant is defined by

$$\varepsilon = \frac{\rho_t}{\rho_i + \rho_e + \rho_t} \; .$$ (5.38)

By introducing the definition of the dielectric constant for electrons, Eq. (5.28), we have

$$\varepsilon = \varepsilon_e - \frac{\rho_i}{\rho_e + \rho_i + \rho_t} \; .$$ (5.39)

From the equation of motion of the ions, Eq. (5.26), we get

$$\frac{\rho_i}{\rho_e + \rho_i + \rho_t} = \frac{\Omega_i^2}{\omega^2} \; ,$$ (5.40)

i.e.

$$\varepsilon = \varepsilon_e - \frac{\Omega_i}{\omega^2} = \varepsilon_e \left(1 - \frac{\omega_q^2}{\omega^2} \right) \; .$$ (5.41)

The Coulomb interaction in the medium is screened so we have

$$V(q, \omega) = \frac{e^2}{\varepsilon_o \, q^2 \, \varepsilon(q, \omega)}$$

$$= \frac{e^2}{\varepsilon_o (q^2 + k_{\rm TF}^2)} \left[1 - \frac{\omega_q^2}{(\omega_q^2 - \omega^2)} \right] \; .$$ (5.42)

As ε can change sign, there is a range of ω for which this interaction can become attractive rather than repulsive. The fact that frequency appears in this interaction means that the interaction is retarded.

It is of interest to calculate the order of magnitude of the attractive part of the potential V, the second term in Eq. (5.42). A dimensionless parameter, $\lambda_{\rm ep}$, which appears in the BCS theory, is the product of the density of states at the Fermi level, $N(E_{\rm F})$, and this attractive potential, V:

$$\lambda_{\rm ep} = N(E_{\rm F}) V \; .$$ (5.43)

As we have seen, a typical phonon wave vector is the Debye wave vector. Substituting all the quantities into Eq. (5.42) $[V_{\rm at}(q_D, 0) = -V]$ we get

$$\lambda_{\rm ep} = \frac{1}{2 + 4.7 a_0 \, (V_0 Z)^{-1/3}} \; .$$ (5.44)

If we take $V_o^{1/3} = 3 \times 10^{-10}$ m and $Z = 1$, we get $\lambda_{ep} \sim 0.33$. For conventional superconductors the experimental value of λ_{ep} is surprisingly close to this order of magnitude estimation (see Table 5.3).

Table 5.3: Values of $\lambda_{ep} = N(E_F)V$ for Al and Nb.

Metal	Calculated λ_{ep}	Experimental λ_{ep}
Al	0.23	0.175
Nb	0.35	0.32

5.7.1. Further insight into the attractive interaction

So far we have only described the simplest theory which shows how an effective attractive interaction can occur due to the motion of ions. If we want to go a step further we have to describe the direct interaction between an electron and the elastic wave propagating into the solid (phonons). An electron propagating in the solid with a wave vector $\mathbf{k}$ can be scattered by a phonon of wave vector $\mathbf{q}$. The phonon disappears and, due to momentum conservation, the electron wave vector becomes $\mathbf{k} + \mathbf{q}$. This is the basic process; the intensity of such a process is measured by what is called the electron-phonon coupling constant. In general, this coupling is small. In the theory of superconductivity, it is the product of this coupling and the density of states at the Fermi level that enters the equations. We denote this product by λ_{ep}; it is a dimensionless coupling parameter.

Are there any other physical properties which depend on this electron phonon interaction? One example is the electron transport. If there is a current in the metal and if electrons are scattered by phonons, the current will decrease. Phonons are excited by the temperature. If the temperature increases, the number of phonons also increases, thus increasing the scattering of electrons and consequently the resistivity. The change of resistivity with temperature is directly related to the electron-phonon coupling. In order to have superconductivity this coupling must be large, hence the room temperature resistivity cannot be small. This explains why the best metals are not superconducting. Pure gold, copper and silver have very small resistivity at room temperature but are not superconducting down to the lowest temperature measured.

5.8. BCS Theory

When the net interaction is attractive, the Fermi sea is unstable with respect to the formation of a bound Cooper pair. Therefore the condensation into pairs will continue until the system reaches an equilibrium point, i.e., until the state of the system has changed so much that the binding energy for yet another pair has become zero. At low temperatures the pairs of electrons can *all be in the same state.* Hence we can say that we have a condensation into such a bound state. As a mathematical treatment is rather difficult we only give an outline of the BCS theory. It relies on a relatively simple model whose main assumptions are as follows:

a) The superconducting ground state can be expressed in terms of Cooper pairs so that the states $(\mathbf{k}, -\mathbf{k})$ are simultaneously occupied or empty.

b) Various interactions in the normal and superconducting states are identical and only the effective screened interaction has to be considered.

c) The effective interaction is zero, except when two electrons of wave vectors $\mathbf{k}$ and $\mathbf{k}'$, have energies close to the Fermi energy. Then the attractive interaction is taken as a constant, $-V$. If ξ_k is the energy measured from the Fermi energy, or more precisely, from the chemical potential μ, then:

$$\xi_k = \varepsilon_k - \mu . \tag{5.45}$$

We impose the condition that, in order for the electrons to attract each other, the energies of both electrons have to satisfy the criterion

$$|\xi_k| \text{ and } |\xi_k'| < k_B \Theta_D \tag{5.46}$$

where Θ_D is the Debye temperature.

This simplified interaction replaces the complicated one which we derived previously and it permits explicit calculations. It turns out that this model can account for most experimental results.

5.8.1. BCS ground state

We first describe the superconducting ground state. We consider always pairs of electrons: one with wave vector $\mathbf{k}$ and the spin up, and the other with wave vector $-\mathbf{k}$ and the spin down. In the normal state, we fill all these pair states up to the Fermi-wave vector $\mathbf{k}_F$. Above $\mathbf{k}_F$ all the pair states are empty. Bardeen, Cooper and Schrieffer found that it is better not to fill the pair states below $\mathbf{k}_F$ completely and have some pairs of electrons in the pair states above $\mathbf{k}_F$. Of course, some kinetic energy is lost as the kinetic energy of the pairs above $\mathbf{k}_F$ has been increased but some potential energy has been gained; as we shall see there is a net gain of energy.

Let v_k be the probability amplitude of the state $(\mathbf{k}, -\mathbf{k})$ which is occupied and u_k the probability amplitude of the empty state; consequently

$$u_k^2 + v_k^2 = 1 . \tag{5.47}$$

The kinetic energy measured from the Fermi energy is

$$E_{\text{kin}} = 2 \sum_k \xi_k v_k^2 . \tag{5.48}$$

The potential energy can be written as

$$E_{\text{pot}} = -V \sum_{kk'} u_k v_k u_{k'} v_{k'} , \tag{5.49}$$

where $-V$ is the attractive potential between two electrons. This potential scatters pairs from state $(\mathbf{k}, -\mathbf{k})$ to state $(\mathbf{k}', -\mathbf{k}')$. This requires the initial state to have the $(\mathbf{k}, -\mathbf{k})$ state occupied and the $(\mathbf{k}', -\mathbf{k}')$ state unoccupied, and vice versa for the final state. The probability amplitude is $u_{k'} v_k$ for the initial and $v_{k'} u_k$ for the final state, thus leading to the above result. The mathematical problem is to minimize the total energy

$$E = E_{\text{kin}} + E_{\text{pot}} \tag{5.50}$$

with respect to the probability amplitude. Here we only quote the result shown in Figure 5.1

$$v_k^2 = \frac{1}{2} \left(1 - \frac{\xi_k}{E_k} \right) , \tag{5.51}$$

where

$$E_k = \sqrt{\xi_k^2 + \Delta^2} \tag{5.52}$$

and

$$\Delta = V \sum_k u_k v_k . \tag{5.53}$$

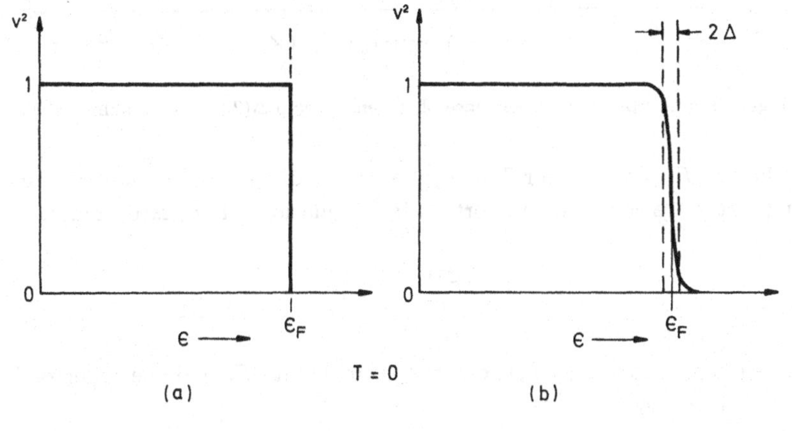

Figure 5.1: Filling of the states **k** in the normal state (a) and in the superconducting state (b).

Δ is a fundamental quantity introduced by Bardeen, Cooper and Schrieffer and is called the *gap* or the superconducting order parameter. It replaces the binding energy introduced by Cooper. Δ depends on the temperature and obeys the relation, known as the *self-consistent BCS equation*

$$\frac{1}{V N(E_F)} = \int_0^{\omega_D} d\xi \, (\xi^2 + \Delta^2)^{-1/2} \tanh \left[\frac{1}{2k_B T} (\xi^2 + \Delta^2)^{1/2} \right] . \qquad (5.54)$$

The temperature dependence of this quantity is shown in Figure 5.2.

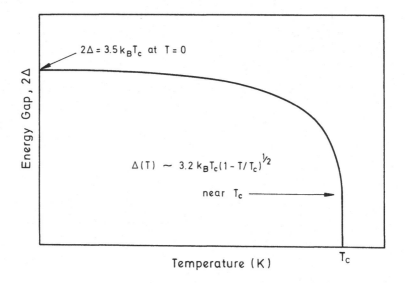

Figure 5.2: Temperature dependence of the energy gap $\Delta(T)$ (after Tinkham 1985).

When $\Delta(T) = 0$, i.e., for $T = T_c$ the energy of the normal and the superconducting states are equal. An important BCS equation relates $\Delta(0)$ and T_c:

$$\frac{2\Delta(0)}{k_B T_c} = 3.52 . \qquad (5.55)$$

The net gain in energy between the normal state E_n and the superconducting state E_s is given by

$$E_n - E_s = \frac{1}{2} N(E_F) \Delta^2 . \qquad (5.56)$$

One should remember that the difference in energy between the two states is given as a function of the critical field $H_c(T)$. Thus we have

$$\frac{1}{2}\mu_0 H_c^2(T) = \frac{1}{2}N(E_F)\Delta^2(T) .$$ (5.57)

Close to T_c we have an approximate formula for $\Delta(T)$:

$$\frac{\Delta(T)}{\Delta(0)} = 1.74\left(1 - \frac{T}{T_c}\right)^{1/2} \text{ for } T \leq T_c .$$ (5.58)

5.8.2. Excitations in the BCS model

In the ground state, all electrons are paired. By breaking the pairs one introduces excitations. This is to be contrasted with a normal metal where excitations are introduced by taking an electron from below the Fermi surface to fill a state above it. If the electron has energy very close to the Fermi surface, this energy difference can be very small. For instance, removing an electron from state $\mathbf{k}$ leaves a hole: the change in energy measured from the chemical potential μ is ξ_k.

If we put this electron into the state $\mathbf{k}'$, it costs an energy $\xi_{k'}$, so the total change of energy is

$$\Delta E_{kk'} = \xi_{k'} - \xi_k = |\xi_k| + |\xi_{k'}|$$ (5.59)

as the state $\mathbf{k}$ is below the Fermi level.

To create an excitation of wave vector $\mathbf{k}$ in a superconductor costs an energy

$$E_k = \sqrt{\xi_k^2 + \Delta^2} .$$ (5.60)

Contrary to the normal state, there is a gap Δ in the excitation spectrum; Δ is temperature-dependent and vanishes at T_c.

When the pair is broken up the ground state is destabilized. The change in energy becomes

$$\Delta E_{kk'} = \sqrt{\xi_k^2 + \Delta^2} + \sqrt{\xi_{k'}^2 + \Delta^2} \geq 2\Delta .$$ (5.61)

There is a minimum change of energy 2Δ required to break up a pair.

Thus, at finite temperatures $(T \neq 0)$ we have both Cooper pairs and single electrons. This appears like a two-fluid model where the superconducting electrons are the pairs and the normal electrons are the excitations made up of single electrons. However, the picture is slightly more complicated as we shall explain in Section 5.8.5 on coherence effects. Nevertheless, such simple way of thinking can be of some help as long as one clearly understands that it may break down completely in certain cases.

Finally, we point out that the density of excitations in the superconductor is given by

$$N(E) = \begin{cases} 0 & \text{for } E < \Delta, \\ N(E_{\rm F}) \frac{E}{\sqrt{E^2 - \Delta^2}} & \text{for } E > \Delta . \end{cases} \tag{5.62}$$

This means that the missing excitations for $E < \Delta$ in the superconducting state are packed up at an energy $E \geq \Delta$. This gives a divergence of the density of states at Δ (see Figure 2.8).

5.8.3. Formula for the critical temperature

BCS theory provides a well-known formula for the critical temperature of a superconductor; it is obtained from Eq. (5.54):

$$k_B T_{\rm c} = 1.13 \hbar \omega_D \, \exp\left(-\frac{1}{\lambda_{\rm ep}}\right) . \tag{5.63}$$

$\lambda_{\rm ep}$ is the dimensionless electron-phonon coupling parameter that we already discussed. Its value for conventional superconductors is very close to ~ 0.3 (see Table 5.3). Here ω_D is the Debye frequency which can be extracted from the contribution of the phonons to the specific heat. This characteristic frequency ω_D varies from one metal to another but only over a small range of values. Instead of ω_D, one can use the Debye temperature Θ_D, defined in Eq. (5.37); Θ_D ranges from 100 K to 500 K. Such a range of Θ_D (and $\lambda_{\rm ep} \sim 0.3$) implies a maximum 'BCS' value of $T_{\rm c} \sim 25$ K.

5.8.4. Specific heat

The behavior of the electronic specific heat of a superconductor is shown in Figure 2.7. As we already mentioned, its behavior in a normal metal at low temperatures is given by

$$C_{\rm en} = \frac{2\pi^2}{3} \, N(E_{\rm F}) \, k_B T . \tag{5.64}$$

At $T_{\rm c}$ there is a drastic change of this behavior: there is a discontinuity in the specific heat

$$\frac{\Delta C}{C_{\rm en}} = 1.43 , \tag{5.65}$$

i.e., the *discontinuity* itself is given by

$$\Delta C = 9.4 \, N(E_{\mathrm{F}}) \, k_B^2 T_{\mathrm{c}} \ . \tag{5.66}$$

At low temperatures, the specific heat decreases exponentially with temperature:

$$C_{\mathrm{es}} \sim \exp\left[-\frac{\Delta(0)}{k_B T}\right] \ . \tag{5.67}$$

Therefore, a measurement of the low temperature specific heat allows us to estimate the gap in the electronic spectrum. It is the existence of the gap which causes this exponential decrease. It would also exist in a semiconductor. This kind of measurement gave the first confirmation of the existence of an energy gap in a superconductor.

5.8.5. Coherence effect

The response of a superconductor to a perturbation can be very different from that of a normal metal. This can be illustrated by considering the effect of an electromagnetic field, but it is much more general. In a normal metal such kind of perturbation scatters an electron from state $\mathbf{k}\uparrow$ to state $\mathbf{k}'\uparrow$.[†] Such scattering is completely independent of the scattering of an electron from state $-\mathbf{k}\downarrow$ to state $-\mathbf{k}'\downarrow$. In a superconductor, this cannot be true as states are occupied by pairs $\mathbf{k}\uparrow$, $-\mathbf{k}\downarrow$. The two transitions are coherent and in quantum mechanics one has to add the probability amplitudes of the two events, i.e., one has the interference effect.

In a two fluid model, the number of normal electrons decreases below T_{c} and hence the possibility of scattering decreases. Consider, for instance, the attenuation of sound in a metal. This attenuation is due to the scattering of elastic waves or phonons by electrons of the metal. If the number of normal electrons decreases, the scattering decreases with the attenuation of sound. Thus one would expect that all such properties, as in ultrasonic attenuation, decrease below T_{c}. Indeed, this has been observed. However, for some properties one may observe an increase below T_{c} followed by a decrease to zero due to the disappearance of normal electrons. This is the effect of interference between the two scattering events.

One of the greatest successes of the BCS theory was the explanation of nuclear magnetic relaxation. Nuclear spins relax in metals because the conduction electrons are able to flip their magnetic moments. Thus one would expect that below T_{c} the time of relaxation of nuclear spins would increase because of the lack of normal electrons. Experimentally it first decreases before increasing. This is the direct effect

[†]The symbol $\uparrow$ denotes 'spin up' and correspondingly $\downarrow$ 'spin down'.

of the coherence between the two scattering processes. This type of experiment was one of the best tests of the pairing theory.

5.9. BCS Theory and Ginzburg-Landau Theory

The Ginzburg-Landau (GL) theory can be derived from the microscopic BCS theory. We are interested to learn how the phenomenological coefficients introduced by Ginzburg and Landau are related to the quantities introduced in the microscopic theory. Instead of giving the values of these coefficients, we will give the values of the quantities derived from these coefficients which have important physical meaning. The first one is the coherence length. The microscopic theory introduces what is known as the BCS *coherence length* ξ_0 which is given by

$$\xi_0 = 0.18 \frac{\hbar v_{\mathrm{F}}}{k_B T_{\mathrm{c}}} , \tag{5.68}$$

where v_{F} is the Fermi velocity, of the order of 10^6 m/s in most metals. For $T_{\mathrm{c}} \sim$ 10 K, $\xi_0 \sim 1800$ Å. This length is temperature-independent. The relation to the temperature dependent coherence length $\xi(T)$ of the GL theory, depends on the mean free path l_e of the electrons. We give only the two limiting cases. For a clean material where the mean free path is much greater than the BCS coherence length one has

$$\xi(T) = 0.74\xi_0 \left(\frac{T_{\mathrm{c}}}{T_{\mathrm{c}} - T}\right)^{1/2} , \quad l_e \gg \xi_0 , \tag{5.69}$$

whereas in the opposite (or 'dirty') limit one has

$$\xi(T) = 0.85 \sqrt{\xi_0 l_e} \left(\frac{T_{\mathrm{c}}}{T_{\mathrm{c}} - T}\right)^{1/2} , \quad l_e \ll \xi_0 . \tag{5.70}$$

This has an important physical consequence: when one alloys superconductors with non-magnetic impurities, one in general does not change T_c and v_{F} very much, hence ξ_0 is rather insensitive to alloying. However, by decreasing the mean free path by alloying one does decrease the GL coherence length $\xi(T)$.

Now let us consider the penetration depth. First we introduce the London penetration depth, $\lambda_L = \left(\frac{m}{\mu_0 n e^2}\right)^{\frac{1}{2}}$. The penetration depth in the Ginzburg-Landau domain is given in two limiting cases by

$$\lambda(T) = \frac{1}{\sqrt{2}} \lambda_L \left(\frac{T_{\mathrm{c}}}{T_{\mathrm{c}} - T}\right)^{1/2} , \qquad l_e \gg \xi_0 , \tag{5.71}$$

$$\lambda(T) = 0.64\lambda_L \sqrt{\frac{\xi_0}{l_e}} \left(\frac{T_c}{T_c - T}\right)^{1/2}, \quad l_e \ll \xi_0 . \tag{5.72}$$

Contrary to the coherence length, which decreases as the mean free path becomes shorter by alloying, the penetration depth increases . The Ginzburg-Landau parameter $\kappa = \frac{\lambda}{\xi}$ is given in the two limits by

$$\kappa = 0.96 \frac{\lambda_L}{\xi_0}, \quad l_e \gg \xi_0 , \tag{5.73}$$

$$\kappa = 0.715 \frac{\lambda_L}{l_e}, \quad l_e \ll \xi_0 . \tag{5.74}$$

κ increases with decreasing mean free path. One can produce a type-II superconductor by alloying a type-I superconductor. These results can be expressed in another way: instead of l_e and λ_L one can use directly measurable quantities of the normal state . The mean free path can be estimated from the electrical resistivity

$$\rho^{-1} = \frac{2}{3} N(E_F) e^2 v_F l_e , \tag{5.75}$$

while λ_L can also be expressed in the form

$$\lambda_L = \left(\frac{3}{2N(E_F)v_F^2 e^2}\right)^{1/2} . \tag{5.76}$$

$N(E_F)$, given in both formulae, can be estimated from the low temperature electronic specific heat.

5.10. Optical Properties of Superconductors

Optical properties of a medium are given in terms of the conductivity and dielectric constant. It is sometimes easier to introduce complex conductivity and discuss these properties in terms of the real and the imaginary parts, σ_1 and σ_2, of the complex conductivity σ:

$$\sigma(\omega) = \sigma_1(\omega) + i\sigma_2(\omega) . \tag{5.77}$$

For optical properties we need to know $\sigma_1(\omega)$ and $\sigma_2(\omega)$ for the range of optical frequencies. This has been calculated within the BCS model and here we only quote the result.

At zero temperature, the real part of the conductivity, $\sigma_1(\omega)$, which is related to dissipation, has a delta-function peak at zero frequency. This corresponds to an infinite conductivity. Subsequently $\sigma_1(\omega)$ is zero for frequencies smaller than $\omega_g = \frac{2\Delta}{\hbar}$. There is no dissipation process in that range of energy as the wave cannot break a pair and there is no single electron to be excited. Above this threshold ('gap') frequency, ω_g, $\sigma_1(\omega)$ increases and for large frequencies becomes the same as in the normal state. This stems from the fact that superconductivity changes only the states of electrons in a range of energy Δ from the Fermi level. For energies large compared with Δ, there is no difference between the normal and superconducting states (see Figure 5.3).

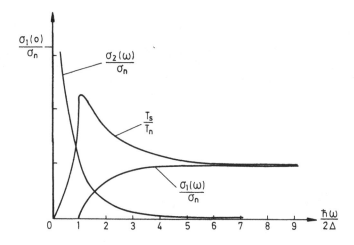

Figure 5.3: Optical conductivity [Eq. (5.77)] of superconductors as a function of frequency (after Lynton 1971).

The behavior of $\sigma_2(\omega)$ is entirely determined by $\sigma_1(\omega)$ and can be analysed by using the Kramers-Kronig relations. There exists a simple formula for the temperature variation of the low-frequency limit of $\sigma_2(\omega)$:

$$\frac{\sigma_2(\omega)}{\sigma_n} = \frac{\pi\Delta}{\hbar\omega} \tanh \frac{\Delta}{2k_BT} \quad \text{if} \quad \hbar\omega \ll 2\Delta \,, \tag{5.78}$$

where σ_n is the conductivity in the normal state.

For $T \ll T_c$

$$\frac{\sigma_2}{\sigma_n} = \frac{\pi\Delta}{\hbar\omega} \,, \tag{5.79}$$

for $T \sim T_c$

$$\frac{\sigma_2}{\sigma_n} = \frac{\pi}{2} \frac{\Delta^2}{k_BT\hbar\omega} \,. \tag{5.80}$$

The interesting region, where superconductors exhibit different behavior from normal metals, is in the range of frequencies near the gap. This range is in the far-infrared where $\lambda \sim 1$ mm. Therefore this range is of interest for applications of superconductors.

Most optical studies are done on the reflectivity of single crystals or on the transmissivity of very thin films. For instance, the transmissivity is given by

$$\frac{T_n}{T_s} = \left\{ \left[T_n^{1/2} + (1 - T_n)^{1/2} \frac{\sigma_1}{\sigma_n} \right]^2 + \left[(1 - T_n)^{1/2} \frac{\sigma_2}{\sigma_n} \right]^2 \right\}^{-1} . \tag{5.81}$$

5.11. Superconductors in Microwave Fields

The mechanism of absorption of radiation of frequency ω at temperature T depends on the ratio $\frac{\hbar\omega}{2\Delta(T)}$. For $\hbar\omega < 2\Delta(T)$, energy can be absorbed by excited particles which are destroyed pairs. This contributes a small resistive component to the surface impedance. For $\hbar\omega > 2\Delta(T)$, such absorption still occurs but, in addition, it is possible to break up pairs. This mechanism leads to a strong increase in absorption as ω becomes greater than $\frac{2\Delta(T)}{\hbar}$.

The general behavior is as follows: at frequencies above $\frac{2\Delta(T=0)}{\hbar}$ there is absorption even at $T = 0$. At frequencies below this threshold there is no absorption at $T = 0$. As T increases the absorption slowly increases initially, while the absorption is due only to excited particles, and then it increases more rapidly as one surpasses the threshold energy $\hbar\omega = 2\Delta(T)$ (see Figure 5.3).

Summary

1. At zero temperature, electrons in a superconductor are paired. In the superconducting state excitations are obtained by breaking up pairs which costs a minimum energy $2\Delta(T)$. The excitation spectrum is given by

$$E_k = \sqrt{\xi_k^2 + \Delta^2} \ .$$

2. The critical temperature of simple conventional superconductors is given by the BCS formula:

$$k_B T_c = 1.13\hbar\omega_D \exp\left[-\frac{1}{\lambda_{\mathrm{ep}}}\right] ,$$

where λ_{ep} is the dimensionless electron-phonon coupling parameter.

3. The difference in energy between the superconducting and normal states is given by

$$\frac{1}{2}\,\mu_0 H_c^2(T) = \frac{1}{2}\,N(E_F)\,\Delta^2(T) \ .$$

4. An important BCS expression relates the zero temperature gap $\Delta(0)$ and the critical temperature:

$$2\Delta(0) = 3.52\,k_B T_c \ .$$

Close to T_c:

$$\Delta(T) = 1.74\,\Delta(0)\left(1 - \frac{T}{T_c}\right)^{1/2} \ .$$

5. The BCS coherence length is defined as

$$\xi_0 = 0.18\,\frac{\hbar v_F}{k_B T_c} \ .$$

6. The characteristic lengths $\xi(T)$ and $\lambda(T)$ introduced in the Ginzburg-Landau theory depend on the mean free path l_e and are related to the BCS coherence length ξ_0: for 'clean' material:

$$\xi(T) = 0.74\,\xi_0\left(\frac{T_c}{T_c - T}\right)^{1/2} , \quad l_e \gg \xi_0 ,$$

for the 'dirty' limit:

$$\xi(T) = 0.85\sqrt{\xi_0 l_e}\left(\frac{T_c}{T_c - T}\right)^{1/2}, \quad l_e \ll \xi_0 .$$

The Ginzburg-Landau penetration depth $\lambda(T)$ in the two limiting cases is given by

$$\lambda(T) = \frac{1}{\sqrt{2}}\lambda_L\left(\frac{T_c}{T_c - T}\right)^{1/2}, \quad l_e \gg \xi_0 ,$$

$$\lambda(T) = 0.64\lambda_L\sqrt{\frac{\xi_0}{l_e}}\left(\frac{T_c}{T_c - T}\right)^{1/2}, \quad l_e \ll \xi_0 .$$

where λ_L is the London penetration depth.

7. A simple description of the superconductor within a two-fluid model can break down due to coherence effects.

8. For time-dependent perturbation, the important frequencies are those related to the gap energy $\hbar\omega \sim \Delta$, i.e., $\nu \sim 10^{11}$ Hz for conventional superconductors. For frequencies much higher than this threshold the superconductor behaves just like a normal metal.

Further Reading

A. A. Abrikosov: *Fundamentals of the Theory of Metals*, North-Holland, Amsterdam, 1988

N. W. Aschcroft and N. D. Mermin: *Solid State Physics*, Holt-Saundres International Editions, 1976

P. G. de Gennes: *Superconductivity of Metals and Alloys*, W.A. Benjamin, New York, 1966

V. L. Ginzburg and D. A. Kirzhnitz (editors): *High Temperature Superconductivity*, Consultants Bureau, Plenum, New York 1982

Ph. Martin and F. Rothen: *Problèmes à n-Corps et Champs Quantiques*, Presses Polytechniques et Universitaires Romandes, Lausanne, 1990

G. Rickayzen: in *Superconductivity* Vol. 1, edited by R.D. Parks, Marcel-Dekker, New York, 1969

J. R. Schrieffer: *Theory of Superconductivity*, Benjamin/Cummings, 1983

M. Tinkham: *Introduction to Superconductivity*, McGraw-Hill, New York, 1975; reprinted by Robert E. Krieger, Malabar, Florida, 1985

Chapter 6.   JOSEPHSON EFFECTS

Preview

We begin with a description of the current-voltage (I-V) characteristics of a tunnel junction between two normal metals, a normal metal and a superconductor, and between two superconductors. In the latter case we present basic physics of the tunnelling of Cooper pairs (Josephson effect). We consider some important principles in Josephson junctions like the Josephson penetration depth, Josephson coupling energy, the hysteretic behavior and the switching time. Tunnel junctions can be replaced by weak links which we define before continuing with a presentation of the basic principles of SQUIDs which represent the most important application of present Josephson technology.

6.1. The Tunnel Effect

Tunnelling is a process by which electrons can travel from one metal, through a narrow vacuum or a thin insulating barrier, into another metal. It is a quantum-mechanical phenomenon. Classically a barrier is impassable if there is insufficient energy available. A ball will not get to the other side of a mountain unless it has sufficient kinetic energy, otherwise it rolls back. Quantum mechanics permits, if the mountain is sufficiently thin, the ball to get to the other side as if there were a tunnel bored through the mountain. Within the quantum world an electron can go through a potential barrier. In what follows we shall discuss several tunnelling phenomena that are important in electronic applications of superconductors.

6.1.1. The NIN junction: Normal metal-insulator-normal metal

Consider a normal metal: the density of states shows a continuous distribution of electronic states. At zero temperature all states below the Fermi energy E_F are completely filled and all the states above E_F are completely empty. The band diagram shown in Figure 6.1 shows the variation of the density of states with energy. $N_1(E)$ and $N_2(E)$ represent the density of states of the metal on the left- and on the right-hand side of the diagram respectively.

From an occupied state at one side of the junction, an electron can tunnel to the other side only if there is an empty state available. Secondly, electrons which tunnel conserve their energies. Therefore, if there is no applied voltage across the junction, there are no empty states available at the other end (of the same energy as the filled initial state) so the electrons cannot tunnel.

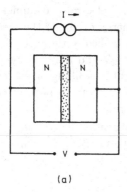

(a)

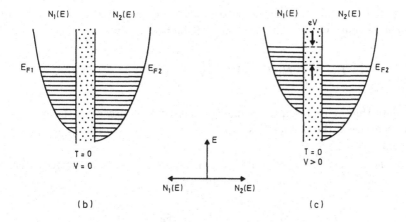

(b)   (c)

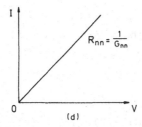

(d)

Figure 6.1: The NIN junction: a) Schematic diagram; b) Density of states vs. energy at $T = 0$ in thermal equilibrium; c) The junction with an applied bias voltage: electrons can tunnel from left to the right; d) I-V curve of an NIN junction.

Obviously we can encourage electrons to tunnel by applying voltage, which is equivalent to an increase in energy of one of the metals relative to the other. For example, if negative voltage is applied to the metal on the left-hand side the energy of all electrons will increase by eV and they will be able to tunnel from metal one to metal two. At finite temperatures, filling of the levels is described by the Fermi function $f(E)$ and tunneling can occur in both directions: the net current is the difference between the two currents.

The number of electrons which can move from left to right in an energy interval dE must be proportional to the number of occupied states on the left and to the number of unoccupied states on the right, i.e., the current from left to right is proportional to

$$N_1(E - eV)f(E - eV)N_2(E)(1 - f(E)) \ .$$

A similar expression can be written for the current flowing from right to left. The net current, which is the difference of the two, is given by

$$I = A \int N_1(E - eV)N_2(E)[(f(E - eV) - f(E)]dE \ . \tag{6.1}$$

Note that A represents the matrix element that describes the probability of transition across the barrier. As it only slowly varies with energy we take it to be constant.

If the density of states is also assumed constant over the range of energy eV then for small V and at low temperatures[†] we get:

$$I = AN_1(E_{\rm F})N_2(E_{\rm F}) eV \tag{6.2}$$
$$= G_{nn} V \ . \tag{6.3}$$

G_{nn} represents the normal state conductance across the junction. The junction behaves in an ohmic way: the current-voltage curve for the NIN junction (see Figure 6.1) is a straight line.

[†]Here we take into account that

$$f(E - eV) - f(E) = -eV \, \frac{df}{dE}$$

and that at low temperatures the derivative of the Fermi function can be approximated by a Dirac delta-function.

6.1.2. The NIS junction: Normal metal-insulator-superconductor: 'The semiconductor model'

We shall now replace the metal on the right-hand side by a superconductor and discuss what happens to the I-V characteristics of the junction within a simple 'semiconductor model'.

Let us recall that in the usual superconductor there is no excitation at low energies. There is an energy gap 2Δ which corresponds to the energy required to break up a Cooper pair. The excitation energies in the BCS model are given by Eq. (5.60):

$$E_k = \sqrt{\xi_k^2 + \Delta^2} \, ,$$

where ξ_k is measured from the Fermi energy.

The density of states of the excitations diverges at Δ and we have

$$N(E) = \frac{EN(E_{\mathrm{F}})}{\sqrt{E^2 - \Delta^2}} \, . \tag{6.4}$$

In order to develop a simple picture of the tunneling, we use a model called *the semiconductor model*. The superconducting state is described as a semiconductor with a gap 2Δ. We extend Eq. (6.4) to the negative values of E and consider all negative energy states as filled. Thus to excite an electron we need an energy 2Δ which is the energy required to break up a Cooper pair.

Once again, at zero voltage electrons cannot tunnel across the barrier as there are no available empty states of the same energy at the other side. Even if we apply a small voltage at $T = 0$, the electrons still cannot tunnel as long as the applied voltage is smaller than the gap, $V < \frac{\Delta}{e}$. As soon as the applied voltage exceeds the gap energy, i.e., $V > \frac{\Delta}{e}$, a large current begins to flow. By writing the expression for the net current, we get a formula similar to the previous case,

$$
\begin{aligned}
I &= A \int_{-\infty}^{+\infty} N_1(E - eV) N_{2s}(E) [(f(E - eV) - f(E)] dE \\
&= \frac{G_{nn}}{e} \int_{-\infty}^{+\infty} \frac{N_{2s}(E)}{N_{2s}(E_{\mathrm{F}})} [(f(E - eV) - f(E)] dE \, .
\end{aligned}
\tag{6.5}
$$

At $T = 0$, by defining

$$G_{ns} = \frac{dI}{dV} \, , \tag{6.6}$$

we get

$$G_{ns} = G_{nn} \frac{N_{2s}(eV)}{N_2(E_{\mathrm{F}})} \, . \tag{6.7}$$

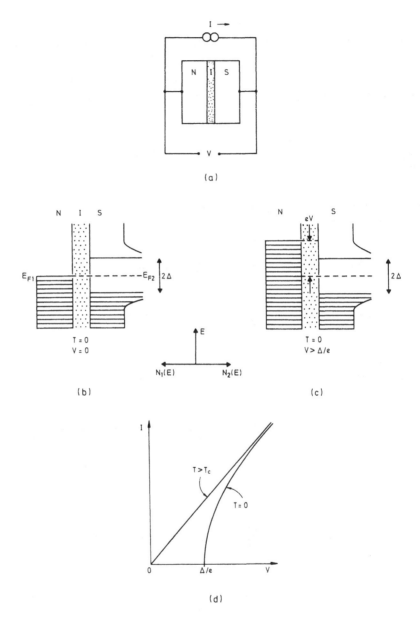

Figure 6.2: The NIS junction: a) Schematic diagram; (b) Density of states vs. energy at $T = 0$ in thermal equilibrium; c) The junction with an applied bias voltage: electrons can tunnel when $V > \frac{\Delta}{e}$; d) *I-V* characteristics of an NIS junction.

At low temperatures the I-V characteristics are strongly non-linear. The conductance G_{ns} measures the density of states of excitations in superconductors at low temperatures. At finite temperatures there are always excited electrons due to thermal excitations and one can measure some current for any voltage. Note also that the resistivity of the junction increases with decreasing temperature.

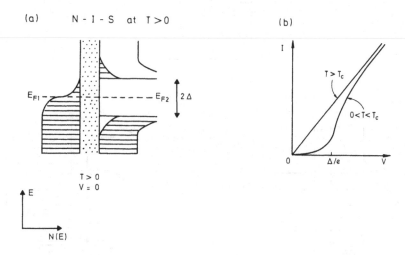

Figure 6.3: Density of states vs. energy of an NIS junction at $T > 0$. b) I-V characteristics for $0 < T < T_c$ and $T > T_c$.

6.1.3. The SIS junction: Superconductor-insulator-superconductor

We now consider tunneling between two superconductors. If the two superconductors are the same, one can easily find that we do not get any tunneling current until we increase the voltage to $\frac{2\Delta}{e}$ (Figure 6.4). At this voltage, we get a discontinuous jump from zero current to a large tunneling current caused by the divergence of the density of states for the empty and filled states. For an asymmetric junction there is a peak in the current at finite temperatures for voltage corresponding to $\frac{|\Delta_1 - \Delta_2|}{e}$ and a large current increase at $\frac{|\Delta_1 + \Delta_2|}{e}$ (see Figure 6.5).

However, in that case the current voltage characteristics is incomplete because at zero voltage it is possible to obtain a current due to Cooper pairs, not to the single particle excitations of the superconductors. This is the Josephson current (see Figure 6.6) and we will deal with it in the next paragraph.

(a)

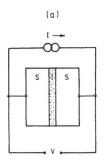

(b)

(c)

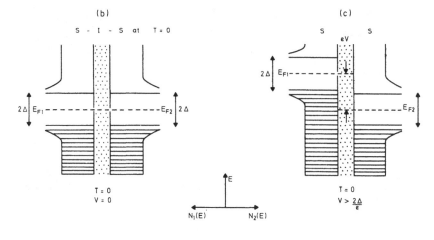

(d)

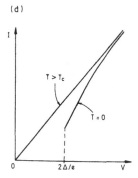

Figure 6.4: The SIS junction: Schematic diagram; b) Density of states vs. energy at $T = 0$ in thermal equilibrium; c) The junction with an applied bias voltage $V > \frac{2\Delta}{e}$; d) the I-V characteristics of an SIS junction for $T = 0$ and $T > T_c$.

(a)

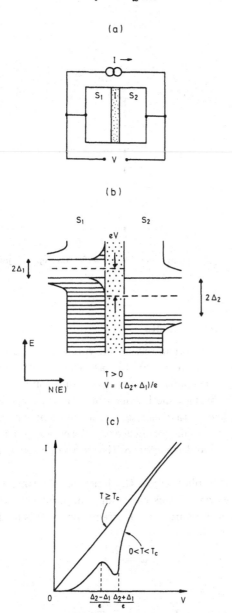

(b)

(c)

Figure 6.5: The asymmetric SIS junction at finite temperatures: a) Schematic diagram; b) Density of states vs. energy with a bias voltage; c) the corresponding I-V characteristics for $T \gtrsim T_c$ and $0 < T < T_c$.

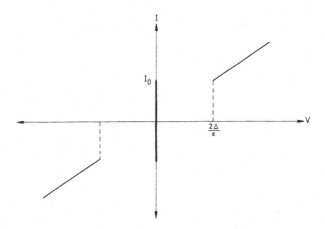

Figure 6.6: Schematic diagram of an *I-V* curve for a Josephson junction at $T = 0$.

6.2. dc Josephson Effect

6.2.1. Various Josephson effects

So far we have seen that the requirement for tunneling is to have an applied voltage across the junction. However, in the case where both metals of the junction are superconducting one encounters an intrinsic superconducting phenomenon: the supercurrent can flow through the barrier from one superconductor into the other even without any applied voltage as long as the supercurrent does not exceed some critical value. This critical current, called the Josephson critical current, is of the order of 1 to 10^3 Acm^{-2} and is very sensitive to the presence of external magnetic field.

When a voltage is applied across the junction, the tunnel current can flow as described above. However, in this case, superimposed onto the dc current, flows an additional ac Josephson current. The frequency of this ac Josephson current is given by the Josephson relation

$$\omega = \frac{2eV}{\hbar} \ .$$

(6.8)

6.2.2. The observation of Josephson effects

Initially Josephson current was observed in a junction made from two superconductors separated by a thin insulating layer, typically ~ 10 Å. Typical normal (state) resistance of such a barrier-layer is $\sim 1\ \Omega$ and the cross-section $\sim 1\ \text{mm}^2$. However, Josephson effects can be observed across any sufficiently localized 'weak link' within a suitable superconducting circuit (see Figure 6.7).

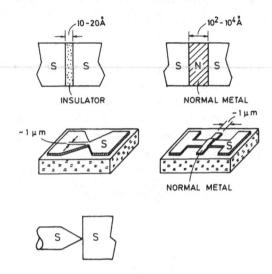

Figure 6.7: Different examples of weak link structures (after Buckel 1977).

Common types of weak link systems are :

i) **SIS junction**: an oxide or other insulator separates two superconducting films. This is the most widely used junction for electronic devices since its preparation can be controlled by thin film fabrication techniques as in semiconductor technology.

ii) **SNS** or **proximity junction**: thin layer of a normal metal evaporated between two superconducting films.

iii) **Point contact** junction between two bulk superconductors. Typically a superconducting wire is ground to a point, the surface is allowed to get oxidized, and then the point is pressed against a piece of bulk superconductor. Point contacts are set up in such a way that the pressure can be adjusted from outside in order to obtain the required current.

iv) Thin film **microbridge** formed by suitable processing of a superconducting film into a narrow constriction.

6.2.3. dc Josephson current

Josephson has shown that the supercurrent which flows between two superconductors separated by a tunnel barrier is related to the phase difference between the order parameters in the two superconductors. The Josephson relation is

$$I = I_0 \sin \gamma \, , \qquad (6.9)$$

where $\gamma = \varphi_2 - \varphi_1$ is the phase difference. φ_i denotes the phase of the corresponding order parameter, $\Psi_i = |\Psi_i| e^{\varphi_i}$ in the superconductor i; with $i = 1$ or 2. This means that if a current I flows across the junction, the phases of the respective order parameters adjust themselves in order to fulfill the Josephson equation (6.9).

Maximum current which can flow across the junction is I_0 and it varies as a function of the temperature dependent gap, $\Delta(T)$:

$$I_0 = \frac{\pi \Delta(T)}{2eR_n} \tanh \frac{2\Delta(T)}{2k_B T} \qquad (6.10)$$

or, at $T = 0$

$$I_0 = \frac{\pi \Delta(0)}{2eR_n} \, . \qquad (6.11)$$

For an oxide junction of 1 mm^2 area, with R_n of the order of 0.1 Ω, I_0 is of the order ~ 10 mA since Δ is of the order 1 meV for a conventional superconductor. The current density through an oxide junction is therefore of the order $\sim 10^4$ A/m^2 which is much smaller than the critical current density of a type-II superconductor. However, for some recently developed junctions, the critical current is higher.

6.2.4. Josephson effect and Ginzburg-Landau equations

Because of the generality of the Josephson relation, we do not attempt to give a general derivation of Eq. (6.9). We will rather show how can one simply derive it from the Ginzburg-Landau equations.

Consider an oxide junction perpendicular to the x direction. The current along this direction is given by the second Ginzburg-Landau relation, Eq. (3.33b). If the oxide layer were infinitely thick, this current would be zero at the surface of the superconductor and we would write

$$-i\hbar \frac{\partial \Psi}{\partial x} - 2eA_x \Psi = 0 \, .$$

If the layer is thin, the current on one side must be related to the superconducting order parameter on the other side. So, phenomenologically, we can write

$$\left(-i\hbar\frac{\partial}{\partial x} - 2eA_x\right)\Psi \text{ (side 1)} = -\frac{i\hbar}{b}\Psi \text{ (side 2)} ,\qquad(6.12)$$

where we have introduced a parameter b that has the dimension of length. The current flowing in side 1 is

$$J_x = \frac{2e}{2m}\left[\Psi_1^*\left(-i\hbar\frac{\partial}{\partial x} - 2eA_x\right)\Psi_1 + \text{c.c.}\right] .$$

Using the boundary condition, Eq. (6.12) we can write for the surface

$$
\begin{aligned}
J_x &= -\frac{i\hbar e}{mb}\left(\Psi_1^*\Psi_2 - \Psi_2^*\Psi_1\right)\\
&= \frac{2e\hbar}{mb}|\Psi_1||\Psi_2|\sin(\varphi_2 - \varphi_1) ,
\end{aligned}
\qquad(6.13)
$$

which is exactly the Josephson relation, Eq. (6.9).

6.3. Effect of a Magnetic Field

The second Ginzburg-Landau equation does not relate the current to the gradient of the phase $\nabla\varphi$ but to the gauge invariant quantity $(\nabla\varphi - \frac{2e}{\hbar}A)$. Thus the gauge invariant phase difference between two superconductors is given by

$$\gamma = \varphi_2 - \varphi_1 - \frac{2e}{\hbar}\int_1^2 \mathbf{A}\cdot d\mathbf{l} .\qquad(6.14)$$

Consider two semi-infinite superconductors separated by an oxide barrier perpendicular to the z-axis, of thickness d in the presence of a magnetic field along the y-axis (as illustrated in Figure 6.8). We can choose for the vector potential the gauge

$$
\begin{aligned}
A_x &= 0,\\
A_y &= 0,\\
A_z &= A_z(x),
\end{aligned}
\qquad(6.15)
$$

and

$$B_y(x) = -\frac{\partial A_z}{\partial x} .\qquad(6.16)$$

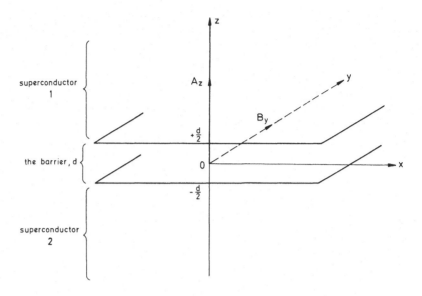

Figure 6.8: Schematic diagram of the field orientation using Cartesian coordinates.

In this case, the gauge invariant phase difference is given by

$$\gamma(x) = \gamma(0) - \frac{2e}{\hbar} A_z(x) (2\lambda + d) . \tag{6.17}$$

The last term is obtained because A_z does not depend on z and is different from zero only in the junction and in each region of thickness λ of the two superconductors on each side of the barrier. The phase difference now varies along the x axis and so does the current.

Consider now the case where the field is uniform across the barrier:

$$B_y(x) = B_0 , \tag{6.18}$$

in which case

$$A_z(x) = -B_0 x$$

and

$$\gamma(x) = \gamma(0) + \frac{2e}{\hbar} B_0 (2\lambda + d)x . \tag{6.19}$$

The current density at point x is given by the Josephson relation, Eq. (6.9)

$$J(x) = J_0 \sin \gamma(x) .$$

It varies sinusoidally along the x axis and the total current is

$$I = L \int_0^D J(x)\, dx \; , \tag{6.20}$$

where L is the dimension of the junction along the y-axis and D, along the x-axis. Hence it can be written as

$$I = LD\, J_0\, \frac{\sin(\pi\Phi/\Phi_0)}{\pi\Phi/\Phi_0}\, \sin\gamma_0 \; . \tag{6.21}$$

LD is the area of the junction and Φ the total flux that threads the junction:

$$\Phi = BD(2\lambda + d) \; .$$

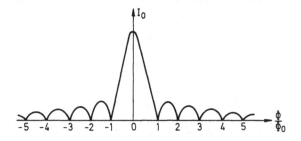

Figure 6.9: The critical current of the junction in the applied magnetic field. It looks like a diffraction pattern of fine single slit in optics ; this is due to the change of phase along the x-axis.

The critical current strongly depends on B as it vanishes for $\Phi = \Phi_0$; Φ_0 is a very small quantity. This result has been verified experimentally providing convincing confirmation of the existence of the Josephson supercurrent.

6.3.1. Josephson penetration depth

So far we have not taken into account the influence of the field created by the Josephson current itself. This neglect has permitted us to take the field as uniform across the junction. If we relax this approximation, a new feature appears: the Josephson current tends to screen the magnetic field out of the junction. There is a Meissner effect for the junction. Indeed the current makes the field $B_y(x)$ dependent on x. We can no longer use a linear relation for $A_z(x)$.

In order to solve the problem, we differentiate $\gamma(x)$ in Eq. (6.17):

$$\frac{\partial \gamma}{\partial x} = -\frac{2e}{\hbar} \frac{\partial A_z(x)}{\partial x} (2\lambda + d)$$

$$= \frac{2e}{\hbar} (2\lambda + d) B_y(x) . \tag{6.22}$$

A Maxwell equation relates the current to curl $\mathbf{B}$ which is given by

$$\text{curl } \mathbf{B} = \left(0, 0, \frac{\partial B}{\partial x}\right) ,$$

thus

$$\mu_0 J_z = \frac{\partial B_y}{\partial x} . \tag{6.23}$$

Differentiating once again, we obtain

$$\frac{\partial^2 \gamma}{\partial x^2} = \frac{2e}{\hbar} (2\lambda + d) \mu_0 J_z$$

$$= \frac{2e}{\hbar} (2\lambda + d) \mu_0 J_0 \sin \gamma . \tag{6.24}$$

Thus we get an equation for γ that we rewrite as

$$\frac{\partial^2 \gamma}{\partial x^2} = \frac{1}{\lambda_J^2} \sin \gamma . \tag{6.25}$$

Here we have introduced a length called the *Josephson penetration depth* given by

$$\frac{1}{\lambda_J^2} = \frac{2e}{\hbar} (2\lambda + d) \mu_0 J_0 = \frac{2\pi (2\lambda + d) \mu_0 J_0}{\Phi_0} . \tag{6.26}$$

If we take $\lambda = 500$ Å, $d = 10$ Å, $J_0 = 10^3$ A/m² we get $\lambda_J = 2$ mm. Typical values of λ_J are of the order of a millimeter. The penetration depth λ_J is for a junction what λ is for a bulk superconductor.

Returning to our equation for γ, we notice that this is the well-known equation of the pendulum. Indeed if we consider a pendulum in a gravitational field, its equation is

$$ml^2 \frac{d^2\theta}{dt^2} = mgl \sin \theta .$$

We can use the analogy (see also Section 6.6.1)

$$x \rightarrow t \ ,$$
$$\gamma \rightarrow \theta \ ,$$
$$\lambda_J^{-2} \rightarrow \omega_0^2 = \frac{g}{l} \ .$$

We have already discussed the case $\lambda_J \rightarrow \infty$ when we neglected the effect of the Josephson current. In that case, the pendulum is whirling around with so much kinetic energy that the gravitational acceleration is negligible. Now if gravity is not negligible, we have two cases: either the pendulum whirls or it oscillates, depending on the initial conditions, i.e., on the total energy of the pendulum. We go from one solution to the other when the total energy of the pendulum is zero, i.e., the kinetic energy is zero when the pendulum is at the top of the circle.

The total energy of the pendulum is zero at $t = 0$ if

$$\left(\frac{d\theta}{dt} \right)_0^2 - 2\omega_0^2 (1 - \cos \theta_0) = 0 \ .$$

By analogy, one can write

$$\left(\frac{d\gamma}{dx} \right)_0^2 - \frac{2}{\lambda_J^2} (1 - \cos \gamma_0) = 0 \ , \tag{6.27}$$

where γ_0 is the phase difference at the edges of the junction and its derivative at the edges of the junction is related to the applied field by Eq. (6.22):

$$\left(\frac{d\gamma}{dx} \right)_0 = \frac{1}{\lambda_J^2 J_0} H \ . \tag{6.28}$$

So we get

$$\cos \gamma_0 = 1 - \frac{1}{2} \left(\frac{H}{\lambda_J J_0} \right)^2 \ . \tag{6.29}$$

As long as the field is small enough for the energy to be negative, one can find an exponentially decreasing solution

$$H(x) \propto \exp \left[-\frac{x}{\lambda_J} \right] \ .$$

The field is screened inside the junction. It will penetrate into the barrier if there is no solution for γ_0 in Eq. (6.29). The highest field which can be screened is that corresponding to $\gamma_0 = \pi$, i.e.,

$$H_{c1}^J = 2\lambda_J J_0 \ . \tag{6.30}$$

This is a very small field. If we take $\lambda_J = 1$ mm and $J_0 = 10^4$ A/m^2, we have $H_{c1}^J = 20$ A/m.

6.4. ac Josephson Effect

We have already pointed out that when dc voltage is applied across the barrier, the supercurrent of Cooper pairs between two superconductors oscillates with a characteristic frequency. This is easily seen from the gauge invariant phase difference

$$\gamma = \varphi_2 - \varphi_1 - \frac{2\pi}{\Phi_0} \int_1^2 A_z \, dz \ .$$

If we differentiate this expression with respect to time, we get

$$\frac{\partial \gamma}{\partial t} = \frac{2\pi}{\Phi_0} \int_1^2 \frac{\partial A_z}{\partial t} \, dz \ . \tag{6.31}$$

A Maxwell equation gives $E_z = -\frac{\partial A_z}{\partial t}$, thus

$$\frac{\partial \gamma}{\partial t} = \frac{2\pi}{\phi_0} \int_1^2 E_z \, dz = \frac{2\pi}{\Phi_0} V \ , \tag{6.32}$$

or

$$\frac{\partial \gamma}{\partial t} = \frac{2e}{\hbar} V \ . \tag{6.33}$$

For a constant voltage V_0, γ is a linear function of time and the Josephson current is

$$J = J_0 \sin \omega_0 t$$

with

$$\omega_0 = \frac{2e}{\hbar} V_0 \ . \tag{6.34}$$

The frequency ω_0 is directly related to the voltage via the fundamental constants e and $\hbar$. For 1 μV the frequency is 484 MHz and the corresponding wavelength is 620 μm.

One can visualize the frequency-voltage relation by saying that the transfer of the Cooper pair from one side to the other requires an energy $2\,eV_0$ which appears as a photon of energy $\hbar\omega_0$.

One may wonder how such a current can be observed. Josephson has proposed the application of both a constant current and a microwave. In that case one observes characteristic steps, called the Shapiro steps, in the I-V curve. Shapiro steps are directly related to the oscillating Josephson current. This follows from the following calculation. Let v and ω be the amplitude and the frequency of the microwave. We have

$$\frac{\partial \gamma}{\partial t} = \frac{2eV_0}{\hbar} + \frac{2ev}{\hbar} \cos(\omega t + \theta) , \qquad (6.35)$$

or

$$\gamma = \frac{2eV_0}{\hbar} t + \frac{2ev}{\hbar \omega} \sin(\omega t + \theta) . \qquad (6.36)$$

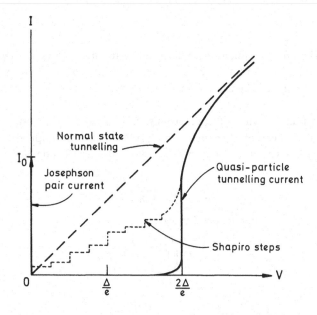

Figure 6.10: Effect of the microwave field on the I-V characteristics (after Tinkham 1985).

In order to express the Josephson current which is superimposed onto the tunnel current, we use the following mathematical expression,

$$\exp\left(ia \sin x\right) = \sum_{n=-\infty}^{+\infty} J_n(a) \exp(inx) , \qquad (6.37)$$

where $J_n(a)$ are Bessel functions of order n. Using this relation the Josephson current can be written as

$$I = I_0 \sum_{n=-\infty}^{+\infty} (-1)^n J_n\left(\frac{2ev}{\hbar \omega}\right) \sin\left[\left(\frac{2eV_0}{\hbar} - n\omega\right) t - n\theta + \gamma_0\right] . \qquad (6.38)$$

The interesting point is that for some values of V_0 the Josephson current becomes a dc current, e.g. whenever the voltage obeys the equation

$$2eV_0 = n\hbar\omega \ . \tag{6.39}$$

For such a voltage, in addition to the tunneling current, there is dc Josephson current which can vary from $-I_0 J_n\left(\frac{2ev}{\hbar\omega}\right)$ to $+ I_0 J_n\left(\frac{2ev}{\hbar\omega}\right)$. Thus, at specific voltages we have current steps in the I-V curve of amplitude

$$2I_0 J_n\left(\frac{2eV}{\hbar\omega}\right) \ . \tag{6.40}$$

In the original experiments of Shapiro, the frequency was 12 GHz producing steps of order $\frac{1}{2}$ mA current at 25 μV voltage separation. The steps depend on the amplitude of the applied microwave in the characteristic way described by the Bessel function.

From the point of view of applications it is interesting to note that the zero-voltage Josephson current is reduced when microwaves are applied. This enables one to detect microwaves. Indeed, at zero voltage, the critical dc Josephson current becomes

$$I_c = I_0 J_0\left(\frac{2ev}{\hbar\omega}\right) \ . \tag{6.41}$$

For small amplitude of microwaves one gets

$$I_c = I_0\left(1 - \frac{2e^2v^2}{\hbar\omega^2}\right) \ . \tag{6.42}$$

6.5. Josephson Coupling Energy

It follows from the Josephson equations that work must be done on the junction to advance the phase from, say, $\varphi = 0$ to $\varphi = \pi$, as we have a voltage and a current. This work is stored as potential energy, the Josephson coupling energy. From Eqs. (6.9) and (6.33) we have

$$E_J = \int JV\,dt = \frac{\hbar I_0}{2e} \int_0^\pi \sin\varphi\,d\varphi = -\frac{\hbar I_0}{2e}\cos\varphi \ . \tag{6.43}$$

The value of this coupling is an important parameter of the Josephson junction.

In order to observe the dc Josephson effect, this energy must be large enough to keep the phases on both sides coupled against thermal fluctuations. If the temperature becomes of the order of this coupling energy, one has fluctuations in the phase difference. The phase coherence across the junction is destroyed and no Josephson current is observed. We define the relevant temperature T_J by

$$k_B T_J = \frac{\hbar I_0}{2e} ,$$ (6.44)

or

$$\frac{T_J}{I_0} = 2.4 \times 10^7 \text{ K/A} .$$ (6.45)

If we operate at temperatures $T > T_J$, the Josephson effect will be 'washed out' by fluctuations and the junction will appear to be normal (see Figure 6.11).

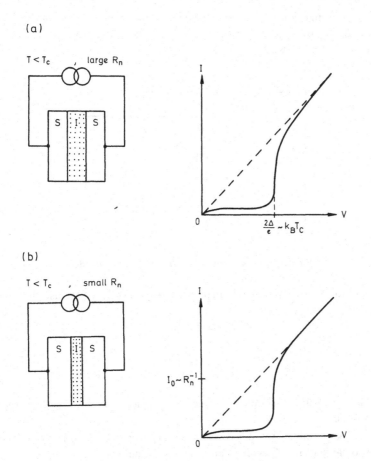

Figure 6.11: Schematic diagram of I-V curve of an SIS junction as a function of normal resistance: a) for large R_n; b) for small R_n (adapted from Wolf 1989).

This equation shows that if one wants to operate a junction at liquid helium temperatures, one needs a value of I_0 greater than 10^{-7} A in order to observe the Josephson current. Thick junctions are unstable against thermal fluctuations as they have large resistance R_n and low critical current I_0 (Eq. (6.11)). In most devices one requires $T/T_J \sim 10^{-1}$–10^{-3} hence at liquid nitrogen temperatures the required value of I_0 can be as large as 3 mA.

6.6. Josephson Junction in a Circuit: The RCSJ Model

For a Josephson junction in a circuit, it is often convenient to use the RCSJ model: Resistively and Capacitively Shunted Junction. The equivalent circuit of the junction is shown in Figure 6.12.

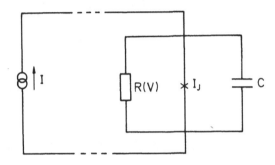

Figure 6.12: Schematic diagram of the RCSJ model.

Indeed, an externally applied current is the sum of three terms: the Josephson current

$$I_J = I_0 \sin \gamma \ , \tag{6.46a}$$

the shunt current

$$I_R = \frac{V}{R(V)} \ , \tag{6.46b}$$

where R is a non-linear shunt resistance which depends on V and it represents junction losses.

Finally, as we have two pieces of metal separated by an insulator, there is a term due to capacitance C which can transport a current

$$I_c = C \frac{dV}{dt} \ . \tag{6.46c}$$

Thus, the externally applied current I is given as

$$I = C\frac{dV}{dt} + \frac{V}{R} + I_0 \sin\gamma \;. \tag{6.47}$$

By using $V = \frac{\hbar}{2e}\frac{d\gamma}{dt}$, we obtain a differential equation for the behavior of the Josephson phase difference γ:

$$I = C\frac{\hbar}{2e}\frac{d^2\gamma}{dt^2} + \frac{\hbar}{2eR}\frac{d\gamma}{dt} + I_0 \sin\gamma \;. \tag{6.48}$$

Due to the highly non-linear character of this equation, one can rarely obtain analytic solutions and often has to resort to numerical methods. However, there is a simple way to qualitatively discuss the solutions of Eq. (6.48): this is the equation of the pendulum so one can use the analogy in order to visualize many of the properties of the junction.

6.6.1. The pendulum analog

Consider a simple rigid pendulum which consists of a light stiff rod of length l with a bob of mass m. The pendulum rotates freely about a pivot (see Figure 6.13). If an external torque T is applied, the pendulum will swing out of the vertical. Let $\theta(t)$ be the angle of deflection at time t. From Newton's law, we have

$$M\frac{d^2\theta}{dt^2} = \text{total torque} \;,$$

where M is the moment of inertia of the pendulum. The total torque is

$$\text{total torque} = T - mgl\,\sin\theta - \eta\frac{d\theta}{dt} \;,$$

i.e., the applied torque plus the torque due to the mass of the bob plus the torque which is due to the viscosity of the air and is opposite to the motion and proportional to the velocity. The Newton's equation is similar to the differential equation for g if we rewrite it as

$$T = M\frac{d^2\theta}{dt^2} + \eta\frac{d\theta}{dt} + mgl\,\sin\theta \;. \tag{6.49}$$

We have an obvious analogy between the two equations (Table 6.1). Visualizing the motion of the pendulum, we can deduce the electrical behavior of a Josephson junction in the following way.

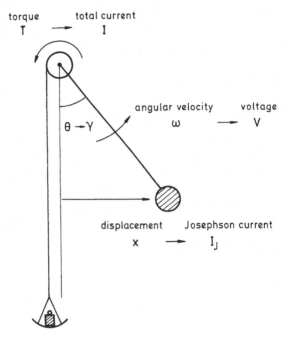

Figure 6.13: Schematic diagram of the pendulum analog; see also Table 6.1 (after Rose-Innes and Rhoderick 1969).

Table 6.1: Corresponding parameters of the Josephson junction and the pendulum (after Rose-Innes and Rhoderick 1969).

Junction	Pendulum
phase difference γ	deflection θ
external current I	applied torque T
capacitance C	moment of inertia M
normal tunneling conductance	viscosity $\eta \dfrac{1}{R}$
Josephson current:	horizontal displacement of a bob:
$I_J = I_0 \sin \gamma$	$x = l \sin \theta$
voltage across the junction:	angular velocity:
$V = \dfrac{\hbar}{2e}\dfrac{\delta \gamma}{\delta t}$	$\omega = \dfrac{d\theta}{dt}$

As an example, we want to study the behavior of the junction when one increases the external current. The equivalent pendulum problem is to study the effect of a torque T applied to the pivot. For a small torque one will have a static deflection θ. If we increase the torque, the angle θ will increase up to $\frac{\pi}{2}$, at which value the pendulum will begin to whirl. Translated to the junction, for a small value of the current we have a dc Josephson current. If we increase the current above a critical value, a voltage develops across the junction.

6.6.2. Hysteretic and non-hysteretic junctions

In the case of a voltage being applied, the dc behavior of a junction does not depend solely on the maximum Josephson current I_0 and does not always show the predicted theoretical behavior that we have calculated for the tunnel characteristics. One can understand the reason in the following way. In the voltage state there is an oscillating Josephson current which interferes with the tunnel current and leads to highly non-linear and complicated behavior. However, there is one case where we can observe the Josephson current and the tunnel branch as we have described. If the capacitance of the junction is large, the oscillating current is short-circuited through it and no ac voltage develops over the junction. Therefore the junction behaves as if there were no ac effect. This is the case if

$$\omega C \gg \frac{1}{R} \, .$$

The characteristic frequency involved, ω_c, is related to the characteristic voltage of the junction, $V_c = \frac{2\Delta}{e}$ by:

$$\omega_c = \frac{2e}{\hbar} V_c \approx \frac{2e}{\hbar} I_0 R_n \, . \tag{6.50}$$

Thus, our condition is

$$\beta_c = \frac{2e}{\hbar} R_n^2 I_0 C \gg 1 \, . \tag{6.51}$$

The parameter β_c that we defined is called the *McCumber parameter*. If β_c is very large, we shall observe the tunnel characteristics and for a given current the junction can be bistable. This bistability is observed experimentally and the junction is *hysteretic* (see Figure 6.14a). Such junctions are mainly used for logic circuits.

If β_c is smaller than one, the behavior is entirely different as strong interference exists. In principle this is not the case of a regular tunnel junction; the order of magnitude shows that $\beta_c \gg 1$. It is possible to shunt the junction, in which case

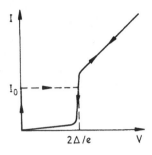

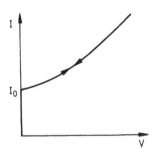

Figure 6.14: a) *I-V* characteristics of a hysteretic junction. b) *I-V* characteristics of a non-hysteretic junction.

the effective resistance is much smaller and β_c can be of order of unity or smaller. In that case the junction is *non-hysteretic* and presents a characteristic which is shown in Figure 6.14b. Such junctions are mainly used in SQUIDs.

6.6.3. The switching speed

It is important to estimate the time it takes for a junction to switch from the Josephson zero voltage state to the voltage state. From the characteristic of the junction, C, V_c and I_0, we can define a time which is an approximation of the switching time τ_s:

$$\tau_s \sim \frac{CV_c}{I_0} \sim R_n C .\tag{6.52}$$

V_c is the voltage at the gap and is fixed for a given material. The speed depends strongly on the thickness of the barrier and decreases for thinner barriers.

To illustrate the speed of a real junction we analyze the characteristics of a PbInAu–PbO–PbBi junction as developed by IBM. V_c is of order 2.5 mV and the critical current is 5×10^3 Acm^{-2} for a junction of thickness $d = 20$ Å.

The capacitance is given by

$$C = \frac{\varepsilon_0 \varepsilon S}{d}, \quad \text{or} \quad \frac{C}{S} = 2.2 \ 10^{-2} \ \text{F/m}^2 = 2.2 \ \mu\text{F/cm}^2 \ ,$$

as ε (PbO) $= 5$ and S is the area of each junction electrode.
From Eq. (6.11) we calculate the area resistance: $SR_n = 0.5 \times 10^{-10}$ Ωm^2.
Hence, the switching time of such a junction is

$$\tau_s \sim R_n C = 1.1 \times 10^{-12} \ \text{s} \sim 1 \ \text{ps} \ .$$

The McCumber parameter β_c is ~ 5.5.

6.7. Weak Links

Weak links are conducting junctions between bulk superconducting samples whose critical current is small compared with those of superconducting electrodes. The term weak links is used to distinguish weakly linked superconducting structures, i.e., those with direct touching electrodes from the tunnel junction that we have considered. Different types of weak links are shown in Figure 6.7.

The first type consists of a normal metal between two superconductors where the thickness of the normal metal film may vary from 100 to 10000 Å. By preparing a narrow bridge of some 1000 Å in superconducting film one can also obtain weak coupling. Point contacts between two superconductors are also weak links. The Josephson effect is observed in all these weak links. The main difference with respect to tunnel junction is that weak links exhibit a non-tunnel type conductivity and a very low capacitance. For various applications of the Josephson effect this is an important advantage over tunnel junctions. Moreover radiation can more easily be coupled in or out of a weak link.

Any weak link is a partition between two superconducting electrodes. Its basic geometrical dimension is its length L. The length is the electrode spacing, i.e., the dimension of the weak link in the direction of the current flow. In a sandwich type of junction this is merely the thickness of the interlayer, but more generally it is the length of the section with changed properties and often is an effective length. Junctions with $L \ll \xi$ will be called short weak links as opposed to long weak links. Strictly speaking, the ideal Josephson effect is observed only in short weak links.

6.7.1. The Aslamazov-Larkin theory

It is possible to prove that for special weak links the Josephson equation holds exactly. It is the case where the temperature T is close to the electrode critical temperature T_c. Then the modulus of the order parameter is small. Consider a weak link made with a normal metal or, to be more specific, another superconductor but above its critical temperature. Below the critical temperature of the electrode, superconductivity is induced by the proximity effect in the normal metal. The order parameter is small and can also be described by a Ginzburg-Landau equation. For the weak links we introduce the coherence length ξ and parameter Ψ_0 given by

$$\xi^2 = \frac{\hbar^2}{2m\alpha} , \quad \Psi_0^2 = \frac{\alpha}{\beta} . \tag{6.53}$$

We notice the change of sign as compared with Eqs. (3.34) and (3.38). Indeed, we are above the critical temperature of the weak link; thus α is positive. We can rewrite the first Ginzburg-Landau equation, Eq. (3.33a) as

$$\xi^2 \left(\boldsymbol{\nabla} - \frac{2ie}{\hbar} \mathbf{A} \right)^2 \Psi + \left(-1 - \frac{|\Psi|^2}{|\Psi_0|^2} \right) \Psi = 0 . \tag{6.54}$$

The boundary conditions are the values of Ψ in the electrodes. We must have

$$\Psi = |\Psi_1| \exp (i\varphi_1) \quad \text{for electrode 1}$$

and

$$\Psi = |\Psi_2| \exp (i\varphi_2) \quad \text{for electrode 2} . \tag{6.55}$$

In the limit where the length of the weak link is sufficiently small as compared with $\xi(T)$ and $\lambda(T)$ the gradient term is the largest one as it is of order $\Psi \left(\frac{\xi}{L} \right)^2$. Thus the Ginzburg-Landau equation reduces to the simple linear Laplace equation

$$\nabla^2 \Psi = 0 . \tag{6.56}$$

The unique solution of this equation with the boundary conditions is given in terms of an unknown function f and it can be written as

$$\Psi = |\Psi_1| e^{i\varphi_1} f + |\Psi_2| e^{i\varphi_2} (1 - f) . \tag{6.57}$$

$f(\mathbf{r})$ satisfies the following boundary value equations

$$\nabla^2 f = 0, \quad \left. \frac{\partial f}{\partial n} \right|_s = 0 , \tag{6.58}$$

where n is normal to the surface.

The current can be expressed using the second Ginzburg-Landau equation, Eq. (3.33b):

$$J_s = C \nabla f |\Psi_1| |\Psi_2| \sin(\varphi_2 - \varphi_1) , \qquad (6.59)$$

where C is a constant. Equation (6.59) has exactly the form of Josephson equation derived in Eq. (6.13).

The fact that the expression for the current in a weak link is the same as the corresponding one for tunnel junctions when $T \to T_c$ suggests that the dc Josephson effect may also take place in an identical manner at arbitrary temperatures. This is however not true. At lower temperatures, the relation between the current and the phase is not of a simple sine function; it is much more complicated. This means that the coupling energy between the two superconducting electrodes does not have the simple form Eq. (6.43) that we assumed for the tunnel junction.

6.8. dc SQUID

We consider a critical current passing through a superconducting ring containing two Josephson junctions in parallel, as illustrated in Figure 6.15. We shall simplify our analysis by assuming that the flux threading each separate junction is negligible. If we further assume that the two junctions have the same critical current I_0 the total current flowing into the parallel circuit is simply given by

$$I = I_0 \left(\sin \gamma_A + \sin \gamma_B \right) , \qquad (6.60)$$

where γ_A and γ_B are the phase differences across junctions A and B respectively:

$$\gamma_A = \varphi_{2A} - \varphi_{1A} , \qquad (6.61a)$$

$$\gamma_B = \varphi_{2B} - \varphi_{1B} . \qquad (6.61b)$$

We assume that the thickness of the wire is greater than λ. Inside the superconductor in circuit C_1 or C_2 the current is zero. At any point we have

$$J_s = 0 , \text{ i.e., } \hbar \nabla \varphi = 2e \, \mathbf{A} ,$$

so we can write

$$\varphi_{1B} - \varphi_{1A} = \frac{2e}{\hbar} \int_{1A}^{1B} A dl , \qquad (6.62a)$$

$$\varphi_{2A} - \varphi_{2B} = \frac{2e}{\hbar} \int_{2B}^{2A} A dl . \qquad (6.62b)$$

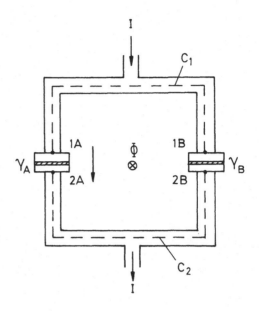

Figure 6.15: Schematic diagram of a dc SQUID.

Summing up these two equations we get

$$\varphi_{1B} - \varphi_{1A} + \varphi_{2A} - \varphi_{2B} = \frac{2e}{\hbar} \oint A dl = 2\pi \frac{\Phi}{\Phi_0} \ , \tag{6.63}$$

as the circulation of A along the closed path $C_1 + C_2$ gives the flux. Thus we have

$$\gamma_A - \gamma_B = 2\pi \frac{\Phi}{\Phi_0} \ . \tag{6.64}$$

So we can write

$$\gamma_A = \gamma_0 + \pi \frac{\Phi}{\Phi_0} \ , \tag{6.65}$$

$$\gamma_B = \gamma_0 - \pi \frac{\Phi}{\Phi_0} \ , \tag{6.66}$$

and the current is

$$I = 2I_0 \sin \gamma_0 \cos \pi \frac{\Phi}{\Phi_0}$$
$$= I_{\max} \sin \gamma_0 \ , \tag{6.67}$$

with

$$I_{\mathrm{max}} = 2I_0 \cos \pi \frac{\Phi}{\Phi_0} \; . \tag{6.68}$$

As long as the current is smaller than I_{max} the phase γ_0 adjusts itself so the current flows without dissipation. I_{max} is the maximum current which can flow in the superconducting interference device (SQUID) without dissipation. Following the properties of the cosine function, I_{max} reaches its maximum for an integral number of flux quanta enclosed within the loop. On the other hand, I_{max} goes to zero whenever the number of enclosed quanta is half-integer. The dependence of the critical current on the enclosed flux provides the basis for the operation of dc SQUID magnetometer, a device that measures minute changes of magnetic flux.

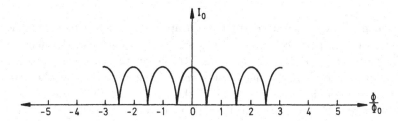

Figure 6.16: Dependence of the maximum supercurrent on flux for a dc SQUID.

If the area of the loop is 1 cm^2 the flux periodicity corresponds to a magnetic field of $B = 2 \times 10^{-11}$ T. In other words, one can resolve flux increments of the order $\sim 10^{-5} \, \Phi_0/\sqrt{\mathrm{Hz}} \sim 2 \times 10^{-20} \, \mathrm{Tm}^2\sqrt{\mathrm{Hz}}$. Obviously such a sensitive flux-measuring device has many potential and actual applications. These we shall briefly discuss in Chapter 8.

6.8.1. Vortex modes

In the above derivation, we have neglected the self-induced flux due to the screening current which is circulating in the loop. The screening current creates a non-negligible flux. In our equation, Φ is the sum of the applied flux Φ_a and the flux Φ_s due to the screening current:

$$\Phi = \Phi_a + \Phi_s \; , \tag{6.69}$$

$$\Phi_s = L I_s \; , \tag{6.70}$$

where L is the self-inductance of the loop. A typical value of L for a loop of radius R, made of a wire of radius a, is

$$L \sim \mu_0 R \ln\left(\frac{R}{a}\right) . \tag{6.71}$$

If the diameter of the loop is a few mm, L is of the order 10^{-9} H.

The screening current is given by

$$I_s = \frac{1}{2} I_0 \left(\sin \gamma_B - \sin \gamma_A\right) \tag{6.72}$$

$$= -\frac{1}{2} I_0 \sin \pi \frac{\Phi}{\Phi_0} \cos \gamma_0 . \tag{6.73}$$

The dependence of critical current on the applied flux is now more complicated. However, it is still periodic in Φ_0. When screening is included, the depth of the periodic modulation of $I_{\max}$ is no longer $2I_0$ [as in the case when screening is neglected, Eq. (6.68)] but a much lower value. Let us note that the screening current I_s reduces the critical current of the SQUID by twice the value of the screening current. Indeed, in each junction we have the current $\frac{I}{2} \pm I_s$ which has to be smaller than I_0. As the maximum screened flux is $\frac{\Phi_0}{2}$, Eq. (6.73) shows that for large values of L the total flux is always close to an integral number of flux quanta. If Φ_a is $\frac{\Phi_0}{2}$, the screened flux is also $\frac{\Phi_0}{2}$ and the screening current is $\frac{\Phi_0}{2L}$. The maximum current is $2I_0 - \frac{\Phi_0}{L}$, rather than $2I_0$. In this strong screening limit, the modulation depth, $\Delta I_{\max}$, is of the order $\frac{\Phi_0}{L} \sim 10^{-6}$ A, regardless of how large I_0 we choose.

Moreover, for Φ_a exactly equal to $\frac{\Phi_0}{2}$, one can either screen completely the field and have $\Phi = n\Phi_0$ with $n = 0$, or the SQUID can pop a full quantum of magnetic flux by setting up a screening current in the opposite direction to keep the magnetic flux within. We have $\Phi = n\Phi_0$ with $n = 1$. The SQUID is bistable. For field Φ_a close to $\frac{\Phi_0}{2}$, we have two stable states which differ by the number of flux quanta stored (see Figure 6.17). The bistability can be used in memory applications for storing one bit of information.

6.8.2. Operation of dc SQUID

In practice one operates with a constant current bias slightly greater than the critical current. The SQUID is always resistive but the voltage across the SQUID is periodic in the magnetic flux with a period of one flux quantum. Roughly speaking,

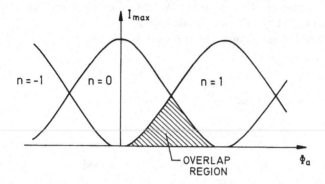

Figure 6.17: Vortex modes of a two-junction SQUID (after Wolf 1978).

the change in voltage across the SQUID due to the applied field equals the change
in its critical current times the resistance and we have approximately

$$\Delta V \sim \frac{R}{L} \Delta \Phi .$$ (6.74)

By measuring the change in voltage one can calculate the change of flux. An
exact calculation can be carried out but such consideration is beyond the scope of
this book.

6.9. rf SQUID

Another variation of the SQUID design is the single point contact junction incorpo-
rated in a loop. In general the entire structure is made of only one superconductor
(see Figure 6.18).

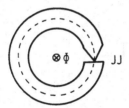

Figure 6.18: Superconducting loop with a single Josephson junction: the rf SQUID.

The flux threading the loop is Φ. We can calculate the phase difference γ in the same manner that we did for the dc SQUID.

As $J_s = 0$ inside the superconductor we have at any point

$$\nabla\varphi = \frac{2e}{\hbar}\mathbf{A} \ . \tag{6.75}$$

By integration over the closed circuit we have

$$\gamma = \varphi_2 - \varphi_1 = -2\pi\frac{\Phi}{\Phi_0} \ . \tag{6.76}$$

The current is given by $I = I_0 \sin\gamma$ and the flux inside the loop is the externally applied flux plus the flux LI where L is the self-inductance of the loop. The current in the loop is always an oscillating function of the flux.

The total current is

$$I = -I_0 \sin 2\pi\frac{\Phi}{\Phi_0} \tag{6.77}$$

with

$$\Phi = \Phi_a + LI \ . \tag{6.78}$$

The magnetic behavior of such a ring depends critically on the value of $\beta = 2\pi\frac{LI_0}{\Phi_0}$, as illustrated in Figure 6.19. If $\beta < 1$, the magnetic behavior is reversible, whereas for $\beta > 1$ the magnetic behavior is hysteretic. In both cases, the magnetic properties are periodic in the externally applied flux with periodicity Φ_0.

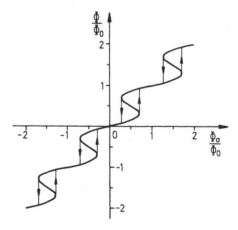

Figure 6.19: The total flux, Φ, as a function of the external flux, Φ_a, for an rf SQUID ($\beta > 1$) (after Cœure 1989).

In the rf SQUID, one uses the hysteretic loop with β between 3 and 6. The applied flux includes both the external flux that is being measured, Φ_p, and an rf flux for biasing. The rf flux is cycled over a range of values including one or more hysteretic loop in Figure 6.19. The resulting dissipation appears as a voltage whose magnitude depends on the flux Φ_p already in the loop. The measured change of voltage is directly related to the change of flux in the device (see Figure 6.20). Such devices can measure only the change of flux, variations of magnetic field or of its gradient.

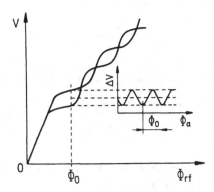

Figure 6.20: Voltage as function of rf bias flux for two extreme values of applied flux Φ_a. Also shown in the figure is the voltage variation for a fixed rf bias flux as the external flux Φ_a is increased (after Cœure 1989).

Summary

1. At low temperatures the *tunnel effect* between a normal metal and a super-conductor allows direct measurements of the density of states of excitations in the superconductor.

2. The *dc Josephson effect* is an intrinsic quantum effect in superconductors: a supercurrent flows through an insulating barrier (or a weak link) without any externally applied voltage. The current is related to the phase difference γ of the order parameters in the two electrodes:

$$I = I_0 \sin \gamma \ .$$

3. At $T = 0$ the *maximum current* which can flow through the Josephson junction is given as

$$I_0 = \frac{\pi \Delta(0)}{2 e R_n} \ .$$

4. The *Josephson coupling* energy is $\frac{\hbar I_0}{2e}$. It has to be greater than $k_B T_c$ in order that the Josephson current may be observed.

5. Within a given circuit the Josephson junction behaves as a *resistively shunted junction with capacitance.*

6. The *McCumber parameter* is defined as

$$\beta_c = \frac{2e}{\hbar} \, R_n^2 \, I_0 C \ .$$

If β_c is large, the junction is *hysteretic*; such junctions are mainly used in logical circuits. If β_c is small the junction is *non-hysteretic*; such junctions are mainly used in SQUIDs.

7. *Weak links are conducting junctions* between bulk superconducting samples whose critical current is small compared with those of the superconducting electrodes. In general, weak links are non-hysteretic.

8. *SQUIDs* are loop-shaped interference devices which exploit the extraordinary sensitivity of Josephson junctions to measure extremely small magnetic flux changes.

Further Reading

A. Barone and G. Paterno: *Physics and Applications of the Josephson Effect*, John Wiley, New York, 1982

James D. Doss: *Engineers' Guide to High-Temperature Superconductivity*, John Wiley, New York, 1989

R. P. Feynman, R. B. Leighton, and M. Sands: *The Feynman Lectures on Physics*, see Chapter 21 in Vol. III, Addison-Wesley, 1966

J. H. Hinken: *Superconductor Electronics*, Springer-Verlag, Berlin, 1989

K. K. Likharev: *Physics of Josephson Junctions*, Addison-Wesley, 1989

T. P. Orlando and K. A. Delin: *Foundations of Applied Superconductivity*, Addison-Wesley, 1991

L. Solymar: *Superconducting Tunneling and Applications*, Chapman and Hall, London, 1972

T. van Duzer and C.W. Turner: *Principles of Superconducting Devices and Circuits*, Edward Arnold, 1982

Chapter 7.   HIGH-TEMPERATURE SUPERCONDUCTING OXIDES

Preview

We present an introduction to the field of high-T_c superconducting (HTSC) oxides. We emphasize several points: their very short coherence lengths that lead to an unusual behavior, particularly in the mixed state, and large anisotropy which can be viewed within a simple model of weakly coupled (super)conducting planes. The macroscopic description requires a generalization of the anisotropic Ginzburg-Landau equations, so we present the Lawrence-Doniach model. Finally we comment on the microscopic origin of high-temperature superconductivity which is still a controversial problem.

7.1. Introduction

Most of the superconducting compounds that we discussed so far were metals. It was reasonable to look for superconductivity in that kind of materials and most research has been done on intermetallic compounds. Some oxide superconductors were known for decades (see Table 7.1), but their transition temperatures were rather small, due mainly to the low numbers of carriers in the metallic state.

Table 7.1: Year of discovery and critical temperature T_c of some oxide superconductors.

Year	Material	T_c/K
1964	NbO	1
1964	TiO	2
1964	doped $SrTiO_{3-x}$	0.7
1965	$K_x WO_3$	6
1966 bronzes	$K_x MoO_3$	4
1969	$K_x ReO_3$	4
1974	$LiTi_2O_4$	13
1975	$Ba(PbBi)O_3$	13
1986	$La_{2-x}Sr_x CuO_4$	38
1987	$YBa_2Cu_3O_7$	92
1988	$Tl_2Ca_2Ba_2Cu_3O_{10}$	125

Two known exceptions were $LiTi_2O_4$ and $BaPbBiO_3$ with critical temperatures of ~ 13 K. This was unusual as their densities of carriers were also very small. The breakthrough came in 1986 when Georg Bednorz and Alex Müller (IBM–Zürich), in their systematic search for new superconductors in metallic Ni- and Cu-oxides, observed an evidence for resistive superconducting transition (with an onset at ~ 30 K) in a fraction of their LaBaCuO sample. This led to the discovery of $La_{2-x}Sr_x CuO_4$

with T_c of ~38 K and subsequently to the widely publicized 'high-T_c revolution' that is described at length in several publications (see reading list at the end of Chapter 1). In this section we will first discuss the most important properties of the $YBa_2Cu_3O_7$ superconductor, initially from a materials scientist's and subsequently from a physicist's point of view. This will enable us to construct a simple model and then discuss the similarities and differences with other high-T_c oxides.

7.1.1. Characteristics of materials

i) High T_c oxides are highly anisotropic, layered structures

Except for some materials (like $Ba_{1-x}K_xBiO_3$), most high-T_c superconducting oxides are cuprate compounds. One of their characteristics is the presence of CuO_2 layers which dominate most properties. If we look at the schematic structure of $YBa_2Cu_3O_6$ presented in Figure 7.1, we immediately notice that it is highly *anisotropic*: The unit cell is developed from that of a tetragonal perovskite tripled along the c-axis and it consists of a sequence of copper-oxygen *layers*.

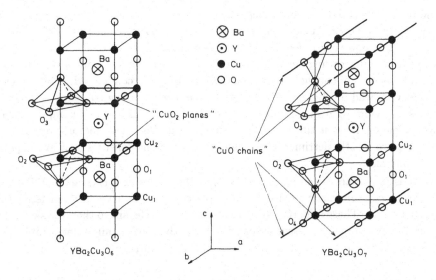

Figure 7.1: Schematic diagrams of (a) $YBa_2Cu_3O_6$ an insulator and (b) $YBa_2Cu_3O_7$ superconducting oxide.

The dimensions of the unit cell are approximately ~ 12 Å and ~ 4 Å in the c- and a- or b-axis directions respectively. An yttrium ion in the center and barium ions above and below the copper-oxygen planes provide the vertical 'spine' of this layered cuprate structure (see Figure 7.1). The fact that the unit cell *consists of layers of copper oxides* will be of great importance for our understanding of the physical properties of these layered structures and will be discussed in more detail in the next section.

ii) Metallic oxides

The second important characteristic of these oxides is their metallic properties. While most oxides are insulating materials, HTSC oxides exhibit metallic behavior. The room temperature conductivities in the a- or b-axis direction of the cuprate crystal are of the same order of magnitude as the conductivities of some disordered metallic alloys. The conductivity is metallic mainly in the CuO_2 planes; perpendicular to these planes, the conductivity is much smaller.

iii) Ceramic materials

The original materials, $La_{2-x}Sr_xCuO_4$ (from now on referred to as **LSCO**) and $YBa_2Cu_3O_7$ (**YBCO**), were synthesized by their discoverers as ceramic pellets. One mixes the correct ratio of constituent oxides, grinds and sinters them, makes a pellet, and following a calcining procedure (at $\sim 950°C$) cools it down in oxygen. Such pellets look very much like pieces of ceramics that we all keep in our hands from time to time during our tea or coffee break (apart from the difference in color). The difference is of course that our tea cups do not superconduct while black pieces of YBCO indeed do (below ~ 92 K). As typical ceramics, high-T_c superconducting oxides also contain grains, grain boundaries, twins, voids and other imperfections. Even some of the best thin films may consist of grains a few microns in diameter; all these are mostly detrimental to high critical current densities that are required for applications.

It is important to emphasize that even the best single crystals of HTSC oxides often contain various defects and imperfections like oxygen vacancies, twins, impurities.... . These imperfections are not only very relevant to their physical properties but possibly even essential for their basic thermodynamic (meta)stability. It may well turn out that various imperfections found in HTSC crystals are intrinsic to these materials.

In general, it is important to understand that the materials science of HTSC oxides is a non-trivial pursuit and that the understanding of phase diagrams, crystal chemistry, preparation and stability of these oxides is still very much in progress. The advancement of our understanding of physics and appearance of applications depend very much on the advancements in materials research. As this book cannot cover more than the basic notions on HTSC oxides, we provide a few reference books on these materials at the end of the chapter.

7.1.2. Characteristic physical properties

i) Superconductors with $T_c \sim 100$ K

This first statement hardly needs much explanation, but we nevertheless emphasize the order of magnitude of T_c. Remember, T_c for Nb_3Ge is 'only' 23 K. We have learned in Chapter 5 that the critical temperature corresponds to the binding energy $\sim k_B T_c$ needed to hold Cooper pairs together in the superconducting state. The fact that $T_c \sim 10^2$ K, i.e. ~ 10 meV, as compared with < 1 meV in conventional superconductors, poses an appealing but profound challenge to theorists interested in the microscopic mechanism of high-T_c superconductivity. We shall briefly discuss this topic at the end of this chapter.

Table 7.2: Critical temperatures of the four most extensively studied high-T_c oxide superconductors (HTSC).

Compound	T_c(K)
$La_{2-x}Sr_xCuO_4$	38
$Y_1Ba_2Cu_3O_7$	92
$Bi_2Ca_2Sr_2Cu_3O_{10}$	110
$Tl_2Ca_2Ba_2Cu_3O_{10}$	125

ii) Quasi-two-dimensional doped insulators

As it stands, the schematic structure of $YBa_2Cu_3O_6$, given in Figure 7.1a, represents an **insulator**. It has to be doped to gradually become a metallic conductor and a superconductor below some critical temperature. The **doping** is achieved by adding additional oxygen which forms CuO 'chains'. These oxygen ions attract electrons from the CuO_2 planes which therefore become metallic (see Figure 7.2).

Note that the correct formula for YBCO material is therefore: $YBa_2Cu_3O_{6+x}$, where x corresponds to partial oxygen content (see Figure 7.2):

for $0.0 < x < 0.4$, $YBa_2Cu_3O_{6+x}$ is an *insulator*,

for $\sim 0.4 < x < 1.0$, $YBa_2Cu_3O_{6+x}$ is a *superconductor*.

The oxygen content can be changed reversibly from 6.0 to 7.0 simply by pumping oxygen in and out of the parallel chains of CuO running along the *b*-axis of

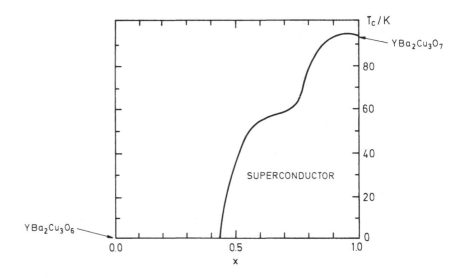

Figure 7.2: Variation of the critical temperature upon doping in $YBa_2Cu_3O_{6+x}$.

Figure 7.1. $YBa_2Cu_3O_6$ is an insulating antiferromagnet. Increasing the oxygen from $O_{6.4}$ makes the crystal metallic, nonmagnetic and superconducting: $T_c = 0$ + first for $O_{6.64}$. The general schematic phase diagram is given in Figure 7.3 as a function of the number of carriers in the CuO_2 plane.

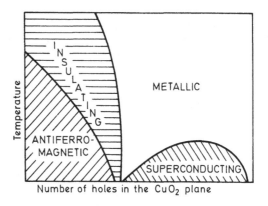

Figure 7.3: Schematic phase diagram of a cuprate superconductor.

iii) Very short coherence length: $\xi \sim 10$ Å

If we recall the BCS-derived formula, Eq. (5.68), $\xi \sim v_F/k_B T_c$, we can immediately expect somewhat shorter coherence lengths in HTSC oxides due to their 10 times higher T_c's. However, due to the low density of carriers, the Fermi velocity in these ionic metals is also lower than in normal metals. This results in a **very short** coherence length, $\xi \sim 10$ Å, which is comparable to the size of the unit cell, and it has profound consequences for the physics of HTSC oxides.

Actually, the coherence length is different for different crystallographic directions and it was experimentally found in YBa$_2$Cu$_3$O$_7$ that ξ_{ab} and ξ_c are ~ 15 Å and ~ 4 Å respectively. Note that ξ_c is roughly equal to the interlayer distance and shorter than the corresponding unit cell length, which clearly poses some conceptual problems.

As we shall see, these remarkably short coherence lengths dominate all material-related properties and cause a rather complex mixed state. Short coherence length also implies that HTSC oxides are **type-II superconductors** with very high upper critical fields B_{c2}.

7.2. Crystal Chemistry of YBa$_2$Cu$_3$O$_{6+x}$

Crystallographers classify the structure of these oxides as of the perovskite type. The name perovskite has no special scientific meaning: it is only a label for a family of structures whose generic class is represented by SrTiO$_3$ and a derived one by K$_2$NiF$_4$ (La$_2$CuO$_4$ structure). The actual name, Perovskite, is a name of a small village in Russia where over the years the crystallographers have found many oxides with similar structures (but of the non-superconducting kind).

YBa$_2$Cu$_3$O$_{6.9}$ has entered in history as the first material with critical temperature above the boiling point of liquid nitrogen (77 K). However, its correct formula reads as YBa$_2$Cu$_3$O$_{6+x}$ and its structure and properties depend on the exact concentration of oxygen. As we have shown in Figures 7.2 and 7.3 it becomes a superconductor for $x > 0.4$. If we carefully observe the unit crystallographic structure in Figure 7.1 we shall notice that the structure consists of a sequence of oxide layers perpendicular to the c-axis as follows:

— Cu-O layer which has two oxygen vacancies as compared with the 'fully oxidized' YBCO perovskite. Cu(1) site in this oxygen layer has coordination 4 and is surrounded by 4 oxygen ions. In YBa$_2$Cu$_3$O$_7$ compound, this is the plane made by the CuO 'chains'.

— Ba-O layer.

— Cu-O layer in which Cu(2) has a coordination number 5 and is surrounded by 5 oxygen ions which form a polyhedra. This is the plane which we call CuO_2 plane.

— Yttrium layer which has 4 oxygen vacancies as compared with the fully oxidized perovskite. The rest of the structure is symmetric with respect to yttrium ion which can be replaced with a whole series of rare earths (except Pr and Yb) without losing superconducting properties.

Copper can be found in two different sites: Cu(1) within CuO_4 'squares' and Cu(2) within a square-based pyramid, CuO_5. The separation of yttrium ions gives the structure a two-dimensional character.

Numerous diffraction studies indicate that most oxygen vacancies occur within planes made of CuO 'chains' ('ab-plane') rather than within the pyramids. In $YBa_2Cu_3O_6$ the chains along the b-axis are oxygen depleted and Cu(1) coordination is only 2 (only two neighboring oxygen ions). This compound is an insulator. By increasing the oxygen concentration one gradually dopes the ab-plane with charge carriers (holes) and it eventually reaches the $YBa_2Cu_3O_7$ composition in which

Table 7.3: Critical temperatures of some HTSC compounds.

Compound	$T_c(K)$
$La_{2-x}M_xCuO_{4-y}$	38
M = Ba,Sr,Ca	
$x \sim 0.15$, y small	
$Nd_{2-x}Ce_xCuO_{4-y}$ (electron doped)	30
$Ba_{1-x}K_xBiO_3$ (isotropic, cubic)	30
$Pb_2Sr_2Y_{1-x}Ca_xCu_3O_8$	70
$\mathbf{R}_1Ba_2Cu_{2+m}O_{6+m}$	
$\mathbf{R}$: Y,La, Nd, Sm, Eu, Ho, Er, Tm, Lu	
$m = 1$ ('123')	92
$m = 1.5$ ('247')	95
$m = 2$ ('124')	82
$Bi_2Sr_2Ca_{n-1}Cu_nO_{2n+4}$	
$n = 1$ ('2201')	$\sim$10
$n = 2$ ('2212')	85
$n = 3$ ('2223')	110
$Tl_2Ba_2Ca_{n-1}Cu_nO_{2n+4}$	
$n = 1$ ('2201')	85
$n = 2$ ('2212')	105
$n = 3$ ('2223')	125

there are no oxygen vacancies. Note that very detailed studies indicate that the maximum in T_c is reached for $x \sim 0.93$ ($T_c = 94$ K) and that for $x = 1.0$ the critical temperature is 'somewhat lower', $T_c = 92$ K. There is also some evidence that the best conduction channel in the normal state is along chains in the b-axis direction (see Figure 7.7).

The optimally doped compound, $YBa_2Cu_3O_{6.9}$, is usually referred to as YBCO or simply as '123'. Its 'average' structure is orthorhombic (remember 'O_6' is tetragonal) but the real material is usually full of various 'defects'. The charge-neutral formula for $YBa_2Cu_3O_7$ can be written as $YBa_2(Cu^{2+})_2(Cu^{3+})(O^{-2})_7$ or as $YBa_2(Cu^{+2})_3(O^{2-})_6(O^-)$.

The exact orthorhombic (quasi-tetragonal) cell dimensions of $YBa_2Cu_3O_7$ are: $a = 3.88$ Å, $b = 3.84$ Å, $c = 11.63$ Å, with a cell volume ~ 173 Å^3.

In $Tl_2Ba_2Ca_{n-1}Cu_nO_{2n+4}$ and the corresponding Bi-compounds, T_c increases with the number of layers of CuO_2. It has been suggested that T_c could increase further for higher n but the compound $n = 4$ seems to have about the same T_c as $n = 3$. The crystallographic structure shows a stacking of planes which for '2212' is $TlO,TlO,BaO,CuO_2,Ca,CuO_2,BaO$ (Figure 7.4d).

We note in Table 7.3 that there is a series of compounds described by the formula $RBa_2Cu_3O_7$, where **R** represents one of the lanthanide elements that can

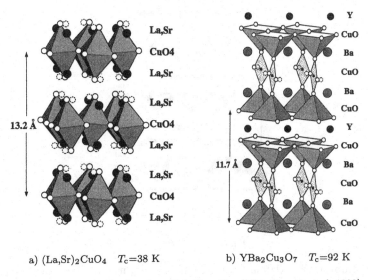

a) $(La,Sr)_2CuO_4$ $T_c=38$ K   b) $YBa_2Cu_3O_7$ $T_c=92$ K

Figure 7.4: Structures of various high-T_c oxides (after Hewatt *et al.* 1989).

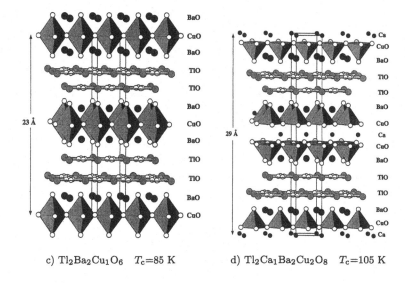

c) $Tl_2Ba_2Cu_1O_6$ $T_c=85$ K

d) $Tl_2Ca_1Ba_2Cu_2O_8$ $T_c=105$ K

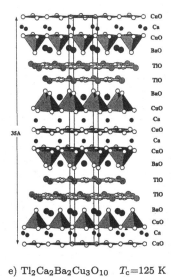

e) $Tl_2Ca_2Ba_2Cu_3O_{10}$ $T_c=125$ K

Figure 7.4: (*Continued*)

replace yttrium in the original '123' structure of Figure 7.1. This means that '123' structure exhibits superconductivity with almost any of the lanthanides. However, an exception is the $PrBa_2Cu_3O_7$ (PrBCO) compound which does not exhibit superconducting properties. This compound effectively behaves as a tunnel barrier in the temperature range where other compounds like YBCO exhibit superconductivity. PrBCO has a matching crystalline structure with YBCO and can be grown between the two $YBa_2Cu_3O_7$ superconducting layers resulting in an artificial 'ingrown' barrier, which is obviously of great potential interest for tunneling devices. Such artificial structures may become very important for tunnel junction technology or for artificial 'construction' of new oxide structures, as will be briefly discussed in Section 5 of Chapter 8.

7.3. Simple Model for Layered Oxides

As shown in Figure 7.5 the structure of YBCO can be schematically represented as a layered structure that consists of two CuO_2 planes separated by Y site. Between these bi-layers are interlayer regions which, in the case of $YBa_2Cu_3O_7$, correspond to the CuO chains.

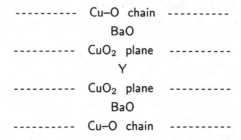

Figure 7.5: Schematic structure of $YBa_2Cu_3O_{6.93}$.

We have mentioned in the previous section that the (super)conductivity essentially takes place within quasi-two-dimensional CuO_2 planes. In the undoped compound, the Cu ions (2+) in this plane are in a d^9 electronic configuration and are antiferromagnetically coupled to other neighboring copper ions, and the plane is insulating. The Cu-O chains can be considered as a 'charge-reservoir' which is needed to transfer the charge into CuO_2 planes. This enables one to consider the HTSC superconductor as CuO_2 planes separated by a charge reservoir (see Figure 7.6). Charge carriers are added by doping: basically by substituting divalent atoms for trivalent ones (like Sr^{2+} for La^{3+} in $La_{2-x}Sr_xCuO_4$) or by adding oxygen to $YBa_2Cu_3O_6$, which enters the compound as O^{2-} and forms CuO chains. To maintain the charge balance, electrons are removed from the copper oxide planes and the remaining holes ('missing electrons') are mobile (hence conduction) and they form 'Cooper pairs' below T_c (hence superconductivity).

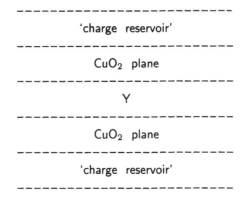

Figure 7.6: A model unit of layered $YBa_2Cu_3O_{6.93}$.

So we can intuitively understand that adding charge carriers from the 'reservoir' into the CuO_2 planes gradually increases the conductivity within the ab-plane. While LBCO has only one 'doped unit-layer' (hence T_c of only ~ 38 K), $YBa_2Cu_3O_7$ has two units separated by Y site while Bi- or Tl- oxides have 1, 2 or 3 (see Table 7.3). In these oxides the role of charge reservoir is evidently played by some other layer like Bi-O rather than CuO as in YBCO. It is interesting to note that, while the conductivity of the CuO_2 planes increases by adding carriers, the superconductivity seems to increase first, reach a maximum for some 'optimal' doping, then decrease and finally vanish for about 0.3 holes per Cu (see Figure 7.3). There is always an optimal doping of the CuO_2 plane which gives the highest T_c. In $YBa_2Cu_3O_{6+x}$ for $x = 0.93$, the highest T_c is ~ 94 K.

In Section 7.7 we will discuss the Lawrence-Doniach model where this initial understanding of layered structure will enable us to understand physical properties in different temperature regimes.

7.4. Normal State of High-T_c Oxides

One of the unusual features of HTSC compounds is the fact that they are very close to an insulating phase. The insulator is slightly doped in order to obtain the superconducting state. If the doping is increased, superconductivity disappears. This rather strange behavior puts forward the problem of the physical properties above the critical temperature. Are they identical to those of the usual metallic state? This is still an open question. Some properties are not similar to those of the conventional metal.

7.4.1. Anisotropy and resistivity

The anisotropy of HTSC oxides is illustrated in Figure 7.7 where we show the temperature dependence of electrical resistivity in a '123' compound. Electrical conductivity in the ab-plane decreases with temperature and is slightly anisotropic: the electrical conductivity is $\sim 30\%$ higher in the b-axis direction due to the CuO chains.

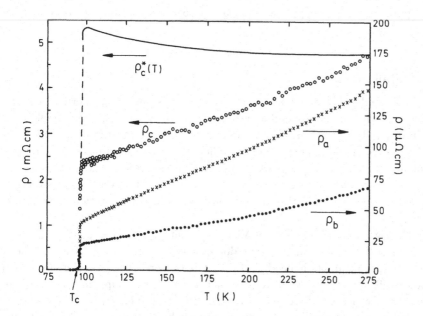

Figure 7.7: Anisotropy of the resistivity of $YBa_2Cu_3O_{6.9}$. [ρ_i ($i = a, b, c$) curves are from Friedmann *et al.* 1990]; $\rho_c^*(T)$ has been measured in oxygen depleted crystals by several groups.

The plane resistivity at ~ 100 K is $\rho_{ab} \sim 60$ $\mu\Omega$cm and it is proportional to the temperature for $T > T_c$. The large anisotropy makes an intrinsic value of ρ_c difficult to measure: several measurements indicate that it is ~ 100 times larger than ρ_{ab} at this temperature. The temperature dependence along c-axis can be metallic for fully doped crystals ($x \sim 0.93$), while semiconducting behaviors (negative slope) are usually observed in oxygen depleted crystals.

We note that the ab-resistivity passes through zero in the best single crystal and film samples and that the temperature dependence of the resistivity is often remarkably linear (up to several hundred Kelvins); this has also been observed in several other HTSC compounds.

The anisotropy of four oxide compounds listed in Table 7.2 rises with increasing T_c. As we shall mainly discuss the properties of $Y_1Ba_2Cu_3O_7$ (YBCO), it is useful to remember that $La_{2-x}Sr_xCuO_4$ exhibits somewhat lower anisotropy, while Bi- and Tl- compounds are much more anisotropic than YBCO. As we mentioned earlier, measured electrical properties of Bi- and Tl- layered compounds are strongly anisotropic in directions parallel and perpendicular to the layers; the ratio is of the order 10^5, i.e., thousand times higher than in '123'! Therefore it is unrealistic in any approximation to neglect the anisotropy of the superconducting properties of these oxides.

7.4.2. Hall number

The Hall number is defined as the inverse Hall constant $1/R_H e$ normalized to unit volume. If the conduction is due either to holes or electrons, as in simple metals, the Hall number gives an estimate of the carrier concentration per unit volume. In YBCO, with the magnetic field parallel to the c-axis and the currents in the ab-plane, the Hall number gives at 300 K a carrier concentration of $\sim 7 \times 10^{21}$ cm^{-3}. This is a very low concentration for a metal. However if one assumes that the carriers are mostly in the CuO_2 planes, then the concentration is close to that of copper.

The sign of the Hall effect of YBCO is positive. This means that the carriers are holes in the CuO_2 planes. When doping increases in $La_{2-x}Sr_xCuO_4$, superconductivity disappears and the Hall effect changes sign. It has been discovered that in some HTCS materials the Hall effect is negative, for example in $Nd_{2-x}Ce_xCuO_4$. The cerium which can be replaced by thorium is 4+, i.e., gives one more electron than lanthanum or neodinium. This electron is believed to go to the CuO_2 planes which become metallic. Thus there are two classes of HTCS oxides: the "p" type where the Hall effect is positive and conductivity in the CuO_2 planes is due to holes, and the "n" type where the Hall effect is negative and conductivity is due to electrons. Of the compounds listed in Table 7.3 only $Nd_{2-x}Ce_xCu_4$ is an "n" type.

7.5. Superconducting State

7.5.1. Characteristic energy ("the gap")

This important, and in principle simple, quantity is very difficult to measure for HTCS materials and a considerable uncertainty exists. The most straightforward method for conventional superconductors is the tunnel effect that is described in Chapter 6. However the tunnel effect is sensitive to a region ξ near the surface.

If the coherence length is several angstroms, one measures only the properties of the surface. In all these materials the surface has probably a different chemical composition. Thus the characteristics obtained are most likely different from those of conventional superconductors. The second direct way to measure Δ is the frequency-dependent conductivity. But here also, the uncertainty is large. One generally measures the gap in units of $k_B T_c$. The BCS value is 3.5. Representative values for the HTCS cuprates seem to be higher and range between 5 and 8 with the characteristic energy of ~ 50 meV.

Moreover the 'gap' is anisotropic. Excitations in the ab-plane have a different gap energy than excitations along the c-axis. Due to difficulties in measuring the gap, one can only say that the gap in the plane is larger than along the c-axis.

7.5.2. Coherence lengths

We have already seen that HTSC oxides with CuO_2 layers are built of superconducting layers separated by dielectric or weakly metallic barriers, all on an atomic scale. Moreover, we have strongly emphasized that a characteristic of these layered crystals is the short coherence length: in YBCO, $\xi_c(0) \sim 4$ Å and $\xi_{ab}(0) \sim 15$ Å. $\xi_c(0)$ is practically equal to the spacing between adjacent conducting CuO_2 planes, which supports the two-dimensional (2D) model with weak coupling in the third dimension between planes (as in our simple model diagram 7.5). In Table 7.4 we list some characteristic coherence lengths. Note that in the much more anisotropic $Bi_2Sr_2Ca_2Cu_3O_{10}$ crystal with $T_c = 110$ K, the estimated $\xi_c(0)$ is only ~ 2 Å (which is remarkably small) while $\xi_{ab}(0) \sim 13$ Å.

Table 7.4: Critical temperature and estimated penetration depth λ_i, coherence length ξ_i, and the upper critical field B_{c2}^i of three HTSC oxides ($i = ab$ or c).

Compound	$T_c(K)$	$\lambda_{ab}(Å)$	$\lambda_c(Å)$	$\xi_{ab}(Å)$	$\xi_c(Å)$	$B_{c2}^{ab}(T)$	$B_{c2}^c(T)$
$La_{2-x}Sr_xCu_4$	38	800	4000	35	7	80	15
$Y_1Ba_2Cu_3O_7$	92	1500	6000	15	4	150	40
$Bi_2Sr_2Ca_2Cu_3O_{10}$	110	2000	10 000	13	2	250	30

7.5.3. Penetration depths

We also list in Table 7.4 the other key length of superconductivity, the penetration depth. As one can see, the value is several thousand angstroms so HTSC oxides belong to extreme type-II superconductors with $\kappa \gg 1$. The large value of λ is related to the low number of carriers as shown in Eq. (3.9).

7.5.4. Critical fields

As we show in Table 7.4, the London penetration depth of YBCO is $\lambda_{ab} \sim 1500$ Å from magnetization measurements, while the coherence length is $\xi_{ab} \sim 15$ Å from critical field measurements. Thus the Ginzburg-Landau parameter, $\kappa_{ab} = \lambda_{ab}/\xi_{ab} \sim 100$, amply satisfies the criterion $\kappa > 1$ for type-II behavior in this compound.

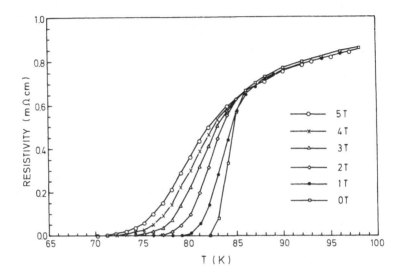

Figure 7.8: Widening of the resistive transition in YBa$_2$Cu$_3$O$_7$ in a magnetic field (A. Shukla *et al.* 1989).

Given the aforementioned crystal structures, the magnetic properties may also be expected to be highly anisotropic. Measurements on single crystals of the LBCO and YBCO systems do indeed show high anisotropy of the critical field. Table 7.4 gives values obtained from resistivity measurements. The estimates of B_{c2} from resistivity data, as shown in Figure 7.8, are inevitably somewhat arbitrary, since the field plotted is that for a stated fraction of the restored normal-state resistivity. It is particularly open to criticism for high-T_c superconductors, in which the resistance-temperature transition curves are especially sensitive to magnetic fields and do not behave as in conventional superconductors. We will describe this behavior in the following section. Moreover, the variation of B_{c2} close to T_c is not linear and the extrapolation to zero temperature relies on a theory whose validity is not known.

Magnetization measurements on YBCO single crystals done by the SQUID magnetometer in magnetic fields up to 5 T (which implies that the temperature range is restricted to within a few degrees of T_c) give substantially higher values than those given in Table 7.4. The slope of the critical field at T_c is very large, of the order: $\frac{B_{c2}^{ab}}{dT}|_{T=T_c} = -10$ T/K and $\frac{B_{c2}^c}{dT}|_{T=T_c} = -1.8$ T/K. The extrapolation to zero temperature ($T = 0$) gives an estimate for $B_{c2}^{ab}(0) \sim 674$ T and $B_{c2}^c(0) \sim 122$ T. Superconductivity in fields up to ~ 100 T has been confirmed in YBCO at 6 K by direct measurements of magnetization in pulsed magnetic fields.

The lower critical fields B_{c1}^{ab} and B_{c1}^c of single-crystal specimens of YBCO have also been estimated from SQUID magnetometer measurements. Again one finds substantial anisotropy with values extrapolated to zero temperature of the order of $B_{c1}^{ab} \sim 2 \times 10^{-2}$ T and $B_{c1}^c \sim 5 \times 10^{-2}$ T. As for B_{c2}, the symbol B_{c1}^{ab} denotes the value of B applied in the ab-plane, and B_{c1}^c of B applied parallel to the c-axis. Note in particular that the anisotropy in B_{c1} has opposite sign to that in $B_{c2} : B_{c1}^c > B_{c1}^{ab}$ but $B_{c2}^c < B_{c2}^{ab}$.

The angular dependence of B_{c2} for $YBa_2Cu_3O_7$ can be well fitted by the anisotropic Ginzburg-Landau model, Eq. (4.15), with a square root ratio of the principal masses equal to 5. In contrast, very sharp angular dependence near $B\|ab$ plane is seen in $Bi_2Sr_2CaCu_2O_{8+y}$. This kind of dependence is found in very thin films of conventional superconductors. The critical field dependence close to $B\|ab$ is very much like that of two isolated CuO_2 planes. This shows clearly the quasi-two-dimensional behavior of this compound.

7.6. Mixed State: Vortex Lattice

7.6.1. Irreversibility line

The mixed state of a high-temperature superconductor reveals a number of unusual features. In a conventional superconductor, the resistive transition shifts downwards with increasing magnetic field. In high-T_c superconductors there is a broadening of the transition (see Figure 7.8). The values of B_{c2} estimated from such a curve are not well defined as they depend strongly on the percentage of the normal resistivity used to define them. The results can be contrasted with diamagnetic B_{c2} which, close to T_c, is much higher and vanishes linearly in a (T_c–T) plot. Thus it is generally believed that the $\rho(T, B)$ curves cannot give a good estimate of B_{c2}.

The second important new feature seems to be the existence of an irreversibility line in the (H, T) phase diagram. This is a non-reversible behavior below a temper-

ature T_i depending on the applied field which manifests itself, for example, in the
following experiment. One first cools the sample without field, and subsequently
a field is applied and the diamagnetic moment is measured at slowly increasing
temperature, obtaining the "zero field cooled" (ZFC) curve. Then one reduces
the temperature back down through the superconducting transition, obtaining the
"field cooled" (FC) curve. A smaller diamagnetic moment is obtained. The zero
field cooled and field cooled curves merge into a common reversible behavior only
above a temperature T_i which is a function of the applied field (see Figure 7.9).
The irreversibility line is found in single crystals and may be determined by several
techniques.

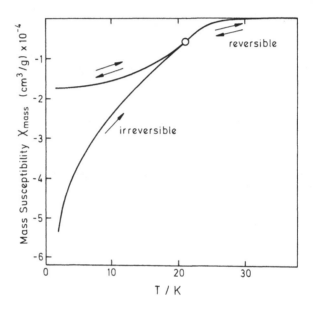

Figure 7.9: Zero field cooled (ZFC) and field cooled (FC) susceptibility of $La_{2-x}Ba_xCuO_4$
(adapted from Müller *et al.* 1987).

A related phenomenon is the large (logarithmic) magnetic relaxation or 'giant
flux creep' observed below the irreversibility line. The large relaxation implies a
rapid approach to the reversible equilibrium state which appears above the irreversibility line (see Figure 7.10).

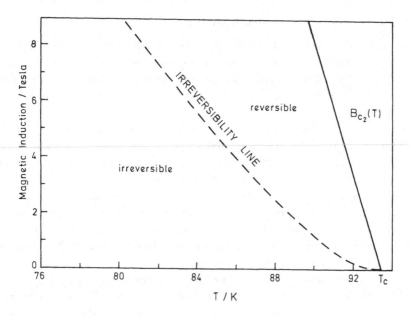

Figure 7.10: Schematic magnetic phase diagram of $YBa_2Cu_3O_7$. Note the irreversibility line (after Worthington *et al.* 1990).

All these phenomena do not seem to be due to disorder in the crystal. These characteristic features exist even in the best quality single crystals, and the increase in disorder does not seem to change very much the position of the irreversible line. This appears to be an intrinsic phenomenon — characteristic of high-T_c materials.

The dissipative mechanism in the mixed state arises from activated flux line motion driven by the Lorentz force. In HTSC materials one expects an important effect due to thermal fluctuations of the flux line lattice. Indeed, as we have already discussed at the end of Chapter 4, the characteristic energy involved in these fluctuations is $\mu_0 H_c^2 \xi^3$ which can be of the order of $k_B T$ as B_c is of the order of ~ 1 T and ξ is very small. Thus these effects have often been attributed to "giant" flux creep.

However, the complete physical picture is more complicated. The explanation strongly depends on the Lorentz force and on the angle between the transport current and the magnetic field due to the angular dependence of Eq. (4.32). This is not always observed. In Bi- and Tl- compounds, the broadening of the transition does not depend on this angle but only on the direction of the magnetic field. The non-observation of the Lorentz force on dissipation has led to many explanations; none of them is fully satisfactory at present.

7.6.2. Observation of the vortex lattice

The vortex lattice has been observed but is somewhat different from that in conventional superconductors. On the scale of a few inter-vortex spacings the pattern is primarily one of hexagonal correlation. On a larger scale, however, the micrographs show the absence of long-range order: the picture is of short-range hexagonal order, not long-range order.

Since high-T_c superconductors are extremely type-II, the inequality $\lambda(T) \gg \xi(T)$ applies, so that the electrodynamics can be described by a local equation of the London type. Since the symmetry is uniaxial, the scalar effective mass used for isotropic superconductors must be replaced by an effective-mass tensor. An elementary consequence of the anisotropic London equation, or of any other description of the vortex structures, is that whereas an isolated vortex in the c-axis direction has circular cross section (ignoring the small difference between the a and b-axes), a vortex in the ab-plane must be elliptical in cross section. This has indeed been confirmed in flux-decoration experiments on YBCO.

A second characteristic is that the vortex lattice is equilateral only for fields parallel to the c-axis. In all other cases the lattice consists of isosceles triangles.

Finally, a third consequence of the anisotropy is that when a magnetic field is applied in a direction intermediate between the c-axis and the ab-plane, the flux lattice, and consequently the specimen magnetization, do not lie in the same direction as the field. Hence the specimen as a whole experiences a torque.

7.7. Mixed State: Critical Currents

One of the most important parameters for potential applications is the critical current and its dependence on the magnetic field. As we have seen, this is not an intrinsic property of the material. In conventional superconductors, the problem is to pin the vortices, and in general increasing the number of defects increases the critical current up to a level which can be close to the de-pairing current. The situation is different in the new ceramics. In a HTCS ceramic, if one measures the critical current, one obtains very low J_c of order hundred A/cm^2 or less. On the contrary, higher J_c is obtained on single crystals. At liquid helium temperature the critical current in the ab-plane is of order 10^7 A/cm^2, and along the c-axis it is 10^5 A/cm^2 (see Figure 7.11). This very unusual situation is mainly due to the very small coherence length of these materials.

To understand the very low value of J_c in a ceramic, we consider it as consisting of small grains, of order of a few microns and of intergrain regions. These intergrain regions, contrary to those in intermetallic compounds, behave like weak links or

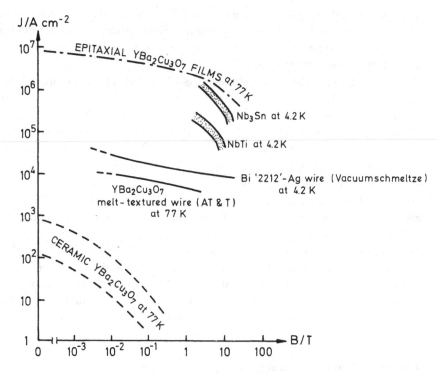

Figure 7.11: Critical current density versus field for several technologically important superconductors (after Evetts 1990).

tunnel junctions. Thus the ceramic is a collection of grains linked by these weak
links or junctions. In order to have zero resistivity, there must be a continuous
superconducting path along the sample. But the critical current of a Josephson
junction or a weak link is very small. Thus with an applied current, these superconducting links are rapidly destroyed and behave as tunnel junctions with high
resistivity. Low values of the critical current of a ceramic reflects the low value of
the critical current of the intergrain regions.

As the critical current of a Josephson junction is also very sensitive to magnetic
field, the critical current of a ceramic decreases rapidly with an applied magnetic
field.

In order to have higher values of J_c, one can eliminate intergrain regions and
use single crystals. Always, due to the small coherence length, some defects in the
crystal behave as weak links and they have to be eliminated in order to have higher
critical currents.

What are the defects which pin vortices in HTCS materials? They have to be of the order of a few angstroms. Contrary to conventional materials, atomic defects can pin vortices. Oxygen vacancies, for instance, are thought to be important pinning centers in these materials.

Measurements of critical current density present considerable experimental difficulties. Contact resistance often restricts the direct method; the alternative is to deduce the critical current density from measurements of magnetic properties, but in such case some assumption has to be made about the current profile within the specimen. In Chapter 4, we have shown that flux creep in HTCS materials can make it difficult to measure the critical current by this method.

All critical current values may be quite uncertain because the measurements are fairly difficult and possibly subject to large reevaluation of the giant flux creep time effects. We give reasonably accepted values of critical currents for comparison with conventional materials.

As a comparison we give the field dependence of the highest J_c measured in HTSC oxides and that of conventional materials, like NbTi and Nb_3Sn (see Figure 7.11). One can easily see that Bi-compounds at 4.2 K and thin films at 77 K offer superior performance to 'old' materials. Relatively high values found in thin-film materials show real promise for commercial use in novel electronic circuits.

7.8. Lawrence-Doniach Model for Layered Oxides

As we emphasized in Section 7.2, HTSC superconductors consist of layers of CuO_2 which alternate with other layers, like Cu-O chains in the case of YBCO 123 phase. The CuO_2 layers are responsible for conductivity and superconductivity while other layers, which donate carriers to the CuO_2 planes, act as either insulating or weakly metallic layers. The density of states at the Fermi energy is very low in these layers. They effectively increase the distance between conducting CuO_2 layers and therefore introduce the observed anisotropy in the crystal.

We have also seen (Chapter 4, Section 4) that a description of superconductivity in an anisotropic material can be given by a simple generalization of the Ginzburg-Landau equations with an anisotropic effective mass: a large one along the c-axis and a small one in the ab-plane. However, this model is appropriate only in compounds with small anisotropy and it becomes invalid in the limit of very large anisotropy. Indeed, the Ginzburg-Landau approach is valid only if the superconducting correlation length is larger than the periodicity of the lattice, $a : \xi > a$.

This approach does not take into account the truly inhomogeneous crystal structure, as one has to average the crystallographic structure within the 'averaged' coherence length. On the other hand such approach is always valid when close to T_c, where the coherence length becomes infinite, $\xi \to \infty$.

When the coherence length at zero temperature is smaller than the lattice parameter or, more precisely, the distance s between CuO_2 planes:

$$\xi_0 = \frac{\hbar v_F}{k_B T} < s \; , \tag{7.1}$$

the anisotropic Ginzburg-Landau approach is valid only if

$$\xi(T) > s \; , \tag{7.2}$$

or

$$\xi_0 \left(\frac{T_c}{T_c - T} \right)^{\frac{1}{2}} > s \tag{7.3}$$

i.e.,

$$\frac{(T_c - T)}{T_c} < \left(\frac{\xi_0}{s} \right)^2 \; . \tag{7.4}$$

The exact expression, which one obtains from detailed theoretical considerations, reads

$$\frac{(T_c - T)}{T_c} < 2 \left(\frac{\xi_0}{s} \right)^2 \; . \tag{7.5}$$

Thus, as long as this condition is not violated, the Ginzburg-Landau approach is valid.

If the ratio ξ_0/s is small, the inequality (7.5) can be violated for temperatures which are not too low. In that case the Ginzburg-Landau approach has to be replaced by a more general model, the Lawrence-Doniach model, that we will now briefly discuss.

In layered compounds, the electron density and the superconducting order parameter are varying in the direction perpendicular to the layers. If the distance between these layers is large compared with ξ_c, the order parameter is very small in between the CuO_2 layers. On the other hand, when ξ_c is large compared with this distance, the inhomogeneity of the electron system is averaged in all macroscopic superconducting properties.

If ξ is small, the order parameter becomes inhomogeneous. As we have seen, it is large within the CuO_2 layers but small between them. This situation is equivalent to that which we discussed in the case of a weak link (see Chapter 6, Section 9). In that case, we describe the order parameter in those regions where it is strong (the electrodes), we ignore the region between and add the Josephson coupling between the two regions (see Figure 7.12). Thus, in this approximation, the layered superconductors can be considered as an array of superconducting layers (the CuO_2 planes) coupled by Josephson interaction.

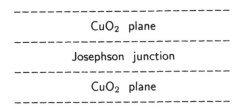

Figure 7.12: Schematic diagram of the Lawrence-Doniach model: two superconducting layers coupled by Josephson junction.

Such a model of layers of atomic thickness coupled by Josephson junctions is often referred to as the Lawrence-Doniach model. In a more precise approach, the free energy functional is written as the sum of the Ginzburg-Landau free energy for the layers and the Josephson coupling energy between layers.

It can easily be shown that if the coherence length along the c-axis becomes large compared with the distance between CuO_2 planes, the model reduces to the anisotropic Ginzburg-Landau model. Thus, the Lawrence-Doniach approach is more general and allows one to treat inhomogeneous geometry. However, we recall that the description of superconductivity within a CuO_2 plane by a Ginzburg-Landau function requires that the order parameter remains small and is therefore not valid down to zero temperature. Nevertheless, this model is very useful and permits one to account for many properties of the highly anisotropic Bi- and Tl- high-T_c compounds. As we have seen earlier, in these compounds the coherence length along the c-axis is much smaller than the distance between the CuO_2 layers.

7.8.1. Upper critical field B_{c2}

The behavior of the upper critical field can be easily understood qualitatively within the Lawrence-Doniach model. If the field is parallel to the c-axis, B_{c2}^c is unaltered, i.e., remains the same as in the Ginzburg-Landau model. For this particular orientation of the field the electrons move within the layers and the phase of the order parameter does not change from one layer to another; therefore the Josephson contribution to the energy does not enter the calculation.

This is not true in the case where the field is parallel to the ab plane: some current will flow between the layers and the phase difference in the order parameter will introduce an energy variation between the layers. Close to T_c the model is reduced to the anisotropic Ginzburg-Landau equation. Equations (4.17) and Eq. (4.18), given in Chapter 4, are valid here.

For the range of temperatures that satisfy the condition

$$\frac{T_c - T}{T_c} > 2 \left(\frac{\xi_0}{s} \right)^2 , \tag{7.6}$$

the field will penetrate between the CuO_2 layers. As λ is large compared with the thickness of the superconducting layers, the field will uniformly penetrate without any cost in magnetic energy. Thus, in this simple model, a thin superconducting film whose thickness is very small compared with λ could support an infinite parallel field without destruction of the superconducting state. This holds only because in Eq. (3.17) we have neglected the decrease of the normal state Gibbs energy with field in a paramagnetic metal. One can show that in this particular case the critical field is given by what is called the paramagnetic Clogston limit. The field B_p is given by

$$B_p = \frac{\Delta}{\mu_B \sqrt{2}} . \tag{7.7}$$

More precisely, the upper critical field B_{c2}^{ab} is given by

$$B_{c2}^{ab} = B_p \sqrt{\frac{T_c - T}{T_c}} \tag{7.8}$$

for

$$\frac{T_c - T}{T_c} > 2 \left(\frac{\xi_0}{s} \right)^2 .$$

This change of behavior between the anisotropic Ginzburg-Landau model close to T_c and the two-dimensional (2D) behavior at lower temperatures, where the superconducting CuO_2 layers are more or less decoupled and behave as independent superconducting planes, can be very easily seen from the curve for $B_{c2}(\theta)$, where θ is the angle between the c-axis and the magnetic field.

Close to T_c and for $\theta = \pi/2$, i.e., for a field close to the ab-plane, one obtains smooth behavior. If the layers are decoupled, one obtains a cusp behavior for $B_c(0)$ around $\theta = \pi/2$, just as in the case of very thin films.

This permits us to determine the characteristic temperature T^* below which the Lawrence-Doniach model gives more appropriate description of layered compounds than the Ginzburg-Landau model. To estimate this temperature T^* for

$Bi_2Sr_2CaCu_2O_{8-\delta}$, we take $T_c = 85$ K, $\xi_0 = 1$ Å and $s = 15$ Å and we obtain

$$T_c - T^* = 0.76 \text{ K} .$$

This indicates that Bi 2212 compound behaves very much as a 2D superconductor (except when very close to T_c) and it can therefore be often approximated as a superconductor-insulator multilayer structure with weak Josephson coupling between the superconducting planes.

7.8.2. Properties of the vortex lattice

We shall now consider the difference between the mixed state of conventional superconductors and the extremely anisotropic high-T_c oxides like Bi and Tl compounds.

Let us suppose that the external field B is parallel to the c-axis of the cuprate crystal. For $B > B_{c1}^c$ the Abrikosov vortex lattice is formed with supercurrents circulating around the cores predominantly located in the CuO_2 planes. As we have learned in Chapter 4, the order parameter ψ vanishes at the center of the vortex core; this means that ψ becomes zero within the superconducting CuO_2 plane. However, as between the planes we are within a Josephson junction region, we cannot use the straightforward description given in Chapter 4.

A better description of this physical picture is to suppose that we have two-dimensional vortex lines in the CuO_2 plane, as in the case of a very thin film. There is an interaction between these 2D vortex lines from one plane to the other and in such a way that one forms flux lines along the c-axis. The difference with respect to the 'conventional' flux lines is that the coupling between these 2D vortices aligned along a line is rather small. This means that the tilt modulus is much smaller than for the 'conventional' vortex line. In fact the tilt modulus C_{44} is reduced by a factor $\frac{m_{ab}}{m_c} \ll 1$. If the anisotropy is large, the tilt modulus becomes much smaller than the shear modulus; this is the opposite to the behavior of 'conventional' lines in the low field limit.

For $B_{c1} \ll B \ll B_{c2}$, one has approximately

$$\frac{C_{44}}{C_{66}} \sim 16\pi \frac{m_{ab}}{m_c} \ll 1 . \tag{7.9}$$

Hence it costs less energy to displace a 2D vortex in layer n with respect to the vortex in an adjacent layer, $n-1$ or $n+1$ (tilt), than with respect to the surrounding vortices within the same layer, n (shear).

In practice, for fields larger than

$$B_{2D} \equiv \frac{\Phi_0}{s^2} \frac{m_{ab}}{m_c} , \tag{7.10}$$

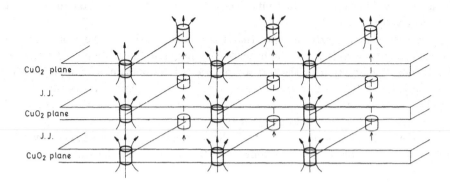

CuO$_2$ plane

J.J.

CuO$_2$ plane

J.J.

CuO$_2$ plane

Figure 7.13: Schematic drawing of two-dimensional (2D) vortex lines in a HTSC oxide.

the pinning and creep of the vortex lattice can be described as in the case of an isolated film. For Bi-2212 compound the field B_{2D} is ~ 0.3 T but ~ 50 T for YBCO.

7.9. Microscopic Theory of High-T_c Superconductivity

7.9.1. Comments on the normal state

One important question which has to be answered before one tries to construct a microscopic theory of HTSC compounds is the nature of the normal state. The simple model, a box of nearly free electrons, which works extraordinarily well in normal metals, does not seem to work so well in HTSC oxides. However, one would like to know whether the usual ('band-picture') theory of metals with some minor changes could explain the unusual properties of HTSC oxides or if one needs a completely new theory to correctly describe their metallic state. As the consensus on such a model does not exist yet, we recall some puzzling experimental results which cannot be explained in a straightforward way within the usual theory of metals and which lead many physicists to believe that the metallic state of cuprate oxides is an entirely new metallic state of matter.

The first such result is that the resistivity in the ab-plane seems to be linear with temperature for an unusually large range of temperatures (up to several hundred Kelvins). At high temperatures the resistivity does not saturate as in common metals; at low temperatures the T^2 term behavior, common to many metals, is generally not observed. The tunnel conductance seems to be linear in voltage in these materials. In conventional materials it is usually independent of voltage at low voltages. The optical properties have also some unusual features: for example, the frequency-dependent conductivity does not follow the simple Drude model that is usually valid for metals.

All these facts clearly indicate that the simple nearly-free electron model is not valid in these oxides. On the other hand, magnetic fluctuations indicate that the Coulomb interaction in these metallic oxides is important as compared with the kinetic energy of electrons and it is not practically negligible as in conventional metals. So at present there remains the challenge of how to describe the normal state of such a correlated anisotropic many-body system (see next section) and of how to devise lucid experiments that will help resolve the theoretical controversies and establish reliable facts concerning both the normal and the superconducting state.

7.9.2. How high can T_c be?

With the discovery of high-temperature superconductors with $T_c \sim 100$ K two obvious questions that most scientists ask are:

i) Can the 'conventional' electron-phonon pairing mechanism account for such a high critical temperature?

ii) Could even higher T_c's be discovered?

While there is no clear consensus among scientists on these topics, most would agree that the role of phonons cannot be excluded (at least) up to about ~ 50 K and that present theories do not exclude the existence of even higher critical temperatures, $T_c > 200$ K (as long as such phases could be thermodynamically stabilized in the laboratory). The following discussion will bring more light into these introductory statements.

The usual BCS formula for T_c, valid for weak electron-phonon coupling, is

$$T_c = \Theta_D e^{-1/\lambda} . \tag{7.11}$$

For conventional superconductors, λ is always of the order of 0.3. Thus, the critical temperatures scale with the Debye temperature. With the exception of Be, the maximum value of Θ_D in metals is of the order of ~ 500 K which implies that a maximum value for T_c is of the order of ~ 25 K.

For stronger electron-phonon coupling McMillan has introduced a more appropriate formula which contains two parameters, λ and μ^*. λ characterizes the electron-phonon interaction as in the BCS case while μ^* takes into account the direct Coulomb repulsion

$$T_c \sim \Theta_D \exp -\frac{1+\lambda}{\lambda - \mu^*} \ . \tag{7.12}$$

For all known materials, $\mu^* \sim 0.13$. Taking as a typical value of the Debye temperature for an alloy, $\Theta_D \sim 300$ K, gives the estimate

$$\lambda \sim 1 \Rightarrow T_c \sim 30 \text{ K} \ .$$

For even stronger electron-phonon coupling, Eq. (7.12) breaks down but for $\lambda = 3$ we can nevertheless give a crude estimate $T_c \sim 75$ K.

There are some serious objections to such a large value for the electron-phonon coupling ($\lambda = 3$). If it did exist, it would probably lead to interesting behaviors in other physical properties (notably resistivity), but this has not been observed. Another possibility is the anomaly of enhanced density of states.

In addition to increasing λ, one may also consider how to increase the exponential prefactor of the BCS theory. Any electronic frequency is usually higher than a phonon frequency as it scales as the square root of the ratio of masses:

$$\omega_e \sim \omega_{\text{ph}} \sqrt{\frac{M}{m}} \ . \tag{7.13}$$

It has been suggested quite some time ago that an electronic mechanism could lead to higher values of the critical temperature. Since the direct interaction between electrons is repulsive, one needs another mediator system which would replace phonons and which would act as follows:

An electron interacts with a system S, which goes into an excited state S^*:

$$e_1 + S \rightarrow e_{1'} + S^* \ . \tag{7.14}$$

A second electron interacts with the system S^*, which therefore returns to its ground state:

$$e_2 + S^* \rightarrow e_2' + S \ . \tag{7.15}$$

The system reverts to its initial state and the electrons are scattered by an effective interaction which is attractive.

Hence we can write

$$T_c = \Delta E e^{-1/\lambda} \ , \tag{7.16}$$

where ΔE is the energy difference between states S and S^* and λ is the coupling between electrons and the system.

In principle, the mediator system can be any physical system as long as it is coupled to the electrons.

In the BCS case, the system S corresponds to the lattice of ions while S^* is the lattice with a phonon propagating through it. However, we can use some other system S; several have been proposed over the years. For instance, magnetic excitations in a magnetic lattice, or electronic excitations of a different system of electrons, etc. Here we will not discuss all these theories; although, in principle, they may well lead to high values of T_c, there is no definite (experimental) confirmation that they are operating in copper oxide planes.

7.9.3. On microscopic mechanisms of high-T_c superconductivity

Most physicists argue that the same microscopic mechanism is operating in all the new oxide superconductors. However there is no general consensus yet on what this mechanism is. Most ideas emphasize the importance of quasi-two-dimensional CuO_2 layers and doping. The main controversy arises with regard to the 'glue', i.e., the origin of the attraction of paired holes which form the condensate (and on the appropriate description of the normal state).

Crudely speaking, we can distinguish four possibilities:

i) A somewhat modified phonon mechanism

It could be larger than expected for two reasons. The density of states could have singularities due to the particular square lattice of the CuO_2 plane, or the electron-phonon coupling could be particularly large for some phonon modes due to certain anharmonic behavior.

ii) Magnetic coupling

This is also a serious possibility as the analog of superconductivity for atoms is the phenomenon of superfluidity. Superfluidity of He^3 is due to the magnetic interactions, and thus to the Coulomb interaction in this material. The analog for electrons is possible.

iii) Electronic coupling

Many years ago, while looking for high-T_c superconductivity, several authors have predicted and searched for superconductivity due to excitons. Excitons are bound electron-hole pairs which exist in semiconductors. This type of excitation could well exist in these cuprates. However, other electronic excitations could also permit superconductivity to exist.

iv) Exotic superconductivity

Very exotic superconductivity has been predicted and linked with the quantum Hall effect. However at present this seems fairly unlikely in these materials.

Some remarks on the BCS phonon-coupling:

- As the electron-phonon coupling is increased, at some stage the crystal is expected to transform to a more rigid and stable structure in which the coupling and T_c are decreased. The critical value of the coupling of harmonic phonons to normal metallic electrons is probably too low to account for the observed T_c's but anharmonic phonons (such as the CuO_6 octahedra breathing mode in a double-well potential) might give an order-of-magnitude enhancement of the electron-phonon coupling without a structural transition.

- The isotope effect, as studied by substitution of ^{18}O for ^{16}O, is weak in these materials. This has prompted the exploration of non-phonon electronic coupling mechanisms. However, a weak isotope effect is not conclusive: there are BCS phonon-assisted superconductors such as Ru and Zr that do not show an isotope effect for reasons that do not apply here. For the exponent α in the BCS expression Eq. (5.1) for the isotope effect, $M^\alpha T_c = $ constant, the substitution of ^{18}O for ^{16}O gives $\alpha \approx 0.02$ in YBCO and $\alpha \approx 0.15$ in LBCO.

Summary

1. HTSC oxides are:

— ionic metals with highly anisotropic, layered structures.

— quasi-two-dimensional doped insulators with an unusual normal state.

— superconductors with $T_c \sim 100$ K, i.e., an order of magnitude higher than for conventional superconductors.

— extreme type-II superconductors with a very short coherence lengths, $\xi \sim 10$ Å and large penetration depths, $\lambda \sim 3000$ Å.

2. Transport properties, characteristic lengths and fields are all highly anisotropic. Upper critical fields can be as high as ~ 150 T and it is possible to achieve $J_c(77$ K$) \sim 10^7$ Acm^{-2} in epitaxial thin films in zero field.

3. Characteristic energy ('gap') is in the range ~ 10 meV, i.e., $\sim 10^{12}$ Hz or ~ 100 μm.

4. Vortex lattice has been observed but the mixed state is rather complex. One can define the 'irreversibility line' below which characteristic magnetic properties are irreversible.

5. Due to the high anisotropy, even the anisotropic Ginzburg-Landau model is not applicable; one can instead use the Lawrence-Doniach model phenomenologically to describe the observed properties.

6. While the conventional phonon mechanism certainly plays some role, the complete microscopic theory is still being developed.

Further Reading

J. G. Bednorz and K. A. Müller (editors): *Earlier and Recent Aspects of Superconductivity*, Springer-Verlag, Berlin, 1989

D. M. Ginsberg (editor): *Physical Properties of High-T_c Superconductors I, II, III*, World Scientific, Singapore, 1989, 1990, 1992

V. L. Ginzburg and D. A. Kirzhnits (editors): *High-Temperature Superconductivity*, Consultants Bureau, London 1982

G. Margaritondo, R. Joynt and M. Onellion (editors): *High-T_c Superconducting Thin Films, Devices and Characterization*, American Institute of Physics, New York, 1989

K. A. Müller and J. G. Bednorz (editors): *Proceedings of High-T_c Workshop in Oberlech*, IBM Journal of R&D, Vol. 33, No. 3, New York, May 1989

A. Narlikar (editor): *Studies of High Temperature Superconductors*, Nova Science Publ., 1989

J. C. Phillips: *Physics of High-T_c Superconductors*, Academic Press, New York, 1989

C. P. Poole Jr., T. Datta and H. A. Farach: *Copper Oxide Superconductors*, Wiley-Interscience, New York, 1988

C. N. R. Rao (editor): *Solid State Chemistry of HTSC Oxides*, Springer, 1990

J. Rowell (editor): *High-temperature Superconductivity*, *Physics Today*, Vol. 44, No. 6, June 1991

'*High-T_c Update*', an international electronic update and database of superconductivity related publications (since 1987).
Editors: Ellen O. Feinberg and John R. Clem, AMES Laboratory, Ames, Iowa 50011-3020, USA
E-mail: FEINBERG@ALISUVAX
Fax: (515)294-1134

Chapter 8.   TECHNOLOGY AND APPLICATIONS

Preview

In this chapter we discuss some applications of superconductors, primarily in high field magnets and superconducting electronics. We briefly discuss conventional Nb-based wire/magnet technology and Nb-based Josephson technology. We review preparation of thin films of HTSC oxides and some of the advances in novel device structures. We conclude with comments on the probable evolution of superconducting technologies.

8.1. Introduction

We have seen in Chapter 4 that high critical current densities needed for potential applications can be achieved only in type-II superconductors. From a practical point of view the most useful aspect of conventional type-II superconductors (e.g. predominantly Nb compounds) is their capacity to carry high transport currents with acceptably low energy dissipation. However, before we briefly discuss some applications of type-II superconductors let us point out that the advancements in materials science, engineering and device technology are well beyond the scope of this textbook. Therefore we list several specialized books at the end of this chapter.

There is a considerable research effort in superconductivity going on around the world; improved materials, technologies and applications are publicized every month. Therefore, we made a deliberate choice to discuss activities predominantly related to thin film applications. While thin films of HTSC oxides may have some impact in passive microwave devices already in the near future, large scale applications of HTSC ceramics, like cables or magnets, will most likely take longer to develop commercially.

8.2. Large Scale and High Current Applications

It is useful to learn 'the rule of thumb' valid for most applications: ideally, the critical temperature T_c of the technological material that one wants to use should be ~ 2 times higher than the temperature of the envisaged application:

$$T_{\text{use}} \sim \frac{T_c}{2} \ . \tag{8.1}$$

This means that superconductors which are used for magnets in liquid helium (4.2 K) should have $T_c > 8$ K. In fact, due to the heating of the magnet, the actual operational temperature is closer to ~ 7 K so one needs $T_c \sim 15$ K. In order to use HTSC oxides in applications at 77 K one should first discover material with

$T_c \sim 150$ K. As such a material has not yet been synthesized, the Tl-compound with $T_c \sim 125$ K is, at present, the best candidate for the 'true' technological material at liquid nitrogen temperature. Note that for most electronics-type of applications one can use the somewhat less stringent rule that the temperature of operation be $\sim \frac{2}{3} T_c$.

In Table 8.1 we present the requirements for critical current densities in a magnetic field for a few selected applications while in Table 8.2 we list a few applications of superconducting coils.

Table 8.1: Requirements for critical current density in a given magnetic field for several applications of superconductors (adapted from Doss 1990).

Application	B(T)	J_c(A/cm^2)
Interconnects	0.1	5×10^6
ac transmission lines	0.2	10^5
dc transmission lines	0.2	2×10^4
SQUIDs	0.1	2×10^2
Motors & generators	~ 4	$\sim 10^4$
Fault current limiters	> 5	$> 10^5$

Table 8.2: Some applications of superconducting coils.

1.	High-field magnets for research in high energy and in condensed matter physics.
2.	Magnetic Resonance Imaging (MRI) which requires extremely uniform magnetic fields of $\sim$1–2 Tesla (formerly known as NMR, nuclear magnetic resonance).
3.	Coils for windings in motors and generators.
4.	Magnetic levitating (MAGLEV) coils for high-speed trains.
5.	Magnetohydrodynamic and electromagnetic thrust systems for propulsion in ships and submarines.

In most of the large scale applications one still uses conventional superconductors like Nb-Ti or Nb$_3$Sn. The corresponding technology of these materials and their applications are discussed at length in specialized books which we list at the end of this Chapter; here we give only a brief summary of some of the most relevant introductory notions.

8.2.1. Materials

Materials used in superconducting magnets must be able to sustain large current densities in high magnetic fields, be strong enough to withstand the stress of the field, and be suitable for manufacturing long wires or tapes. Until now only a few conventional materials, like Nb-Ti, Nb_3Sn and V_3Ga, have been used commercially. The wires made of Chevrel phases or HTSC oxides, mainly YBCO and BSCO, are still under development.

One of the first alloys used in the fabrication of superconducting coils was Nb-Ti. It is a very ductile alloy of body-centered-cubic structure (A2), and can be easily formed into wires. It is commonly used for applications in magnetic fields up to ~ 10 Tesla. At 4.2 K, in a field of ~ 5 T, Nb-Ti alloy can transport current densities of the order of 10^5 Acm^{-2} (see Figure 7.11).

Other materials for high field applications are Nb_3Sn and V_3Ga. These cubic A15 structures exhibit somewhat higher critical temperatures than Nb-Ti, ($T_c = 18.0$ K and 14.8 K respectively) and larger upper critical fields ($B_{c2} = 24$ T and 23 T respectively). While Nb_3Sn is relatively brittle and rather difficult to fabricate into the required conductor-configuration (see the next section), it can be used without quenching in fields > 20 T. At 4.2 K and 10 Tesla, Nb_3Sn can transport current densities in excess of 2×10^5 Acm^{-2}. Other compounds of the A15 family, such as Nb_3Ge with the highest T_c of ~ 23 K, are much more difficult to produce in a wire type of configuration and did not reach commercialization.

As some HTSC oxides have $B_{c2} > 100$ T (at $T = 0$) one would expect to be able to develop magnets of unprecedented field intensity. However, as we have learned in Chapter 7, they are very anisotropic and their critical current densities are still too low in the bulk ceramic form. Even if problems associated with ductility and J_c could be partly solved by addition of Ag, the limitations on the field strength, due to the enormous Lorentz forces involved at higher fields, may be the main factor in limiting the construction of magnets. This, together with difficulties in fabrication of these materials, will somewhat delay the introduction of HTSC oxides into magnet type of applications.

Among numerous Chevrel phases (see Chapter 1), $PbMo_6S_8$, with its extraordinary high upper critical field, $B_{c2} \sim 60$ T, is the most promising material for cable or magnet applications. Despite its brittleness and other intrinsic materials problems which limit J_c, a few groups are currently developing wires of this compound.

Transition-metal carbides or nitrides of the NaCl-type structure (B1), and in particular, the NbN compound are also of technological interest. The latter exhibits large values of J_c and B_{c2} and is being tested on superconducting electronics. The electrodes in tunnel junctions and related devices are nowadays fabricated mainly of NbN or 'pure' Nb. We will learn more about this technology in Section 8.3. Since compounds of $Bi_2Sr_2Ca_2Cu_3O_{10}$ with added Ag are reportedly somewhat more ductile than $YBa_2Cu_3O_{6.9}$, perhaps some variant of these materials might be of use in the construction of magnet wires. Particularly as the suitably processed mixtures of $Bi_2Sr_2Ca_2Cu_3O_{10}$ with Ag exhibit remarkably high critical current densities (at 4.2 K) in high magnetic fields (see their $J_c(B)$ behavior in Figure 7.11).

Table 8.3: Critical temperatures, characteristic lengths and upper critical fields of some technologically important superconductors.

Material	T_c(K)	Penetration depth λ(nm)	Coherence length ξ(nm)	Upper critical field B_{c2}(T)
Nb	9.2	40	38	0.2
NbTi	9.2	60	40	14
NbN	16	250	4	16
Nb_3Sn	18	80	3	24
$YBa_2Cu_3O_7$	92	150/600	1.5/0.4	150/40
$Bi_2Sr_2Ca_2Cu_3O_{10}$	110	200/1000	1.4/0.2	250/30

8.2.2. Wires, cables and magnets

The main problem of wire construction for magnets is the dissipation of energy (usually associated with flux creep) which can lead to disastrous consequences. For example, it can cause *thermal runaway*: the material rapidly heats up and the entire energy stored in the magnet gets converted suddenly into thermal energy. To prevent this, one has to design the wire in such a way that the outflow of heat to the surrounding material is always greater than the rate of dissipation. This requirement for stability implies that the 'wire' should have very small diameter. Typical diameter of NbTi filament is < 0.1 mm, and thinner than 10 μm for Nb_3Sn filaments. Such filaments are then arranged inside a copper matrix (see Figure 8.1). Copper enables one to draw this composite into a wire; it also removes heat which could have been initiated by thermal fluctuations.

We now outline some aspects of the preparation of Nb_3Sn wires as an example of the 'conventional' wire-magnet technology. For improved stability the technological objective is to prepare wire with superconducting Nb_3Sn filaments of a very small diameter (< 10 μm). Furthermore, during the mechanical formation to the final wire diameter, all the constituents have to be ductile: one uses mainly combinations of Nb, Cu, Cu-Sn and Ta.

Figure 8.1: Cross section of multifilamentary Nb_3Sn wire prepared by the 'Internal Sn Diffusion' Technique (after Flükiger 1989).

Fabrication methods of multifilamentary wires of Nb_3Sn are based on the 'Bronze Diffusion Process': the atoms diffuse out of the Cu-13wt.%Sn bronze and react with Nb to form Nb_3Sn. Subsequent processing follows either the 'Bronze Route' or the 'Internal Sn Diffusion' process. The main difference between the two is that the 'Bronze Route' requires several intermediate annealing during the wire drawing process. In the 'Internal Sn Diffusion' process the whole wire deformation

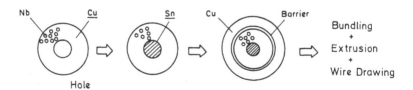

Figure 8.2: Schematic representation of the 'Bronze Route' and of the 'Internal Sn Diffusion' process for producing multifilamentary Nb_3Sn wires (after Flükiger 1989).

procedure can be performed without annealing and is therefore economically more favorable. In the 'Bronze Route' the amount of Sn in the Cu matrix is limited to 13 wt.% due to restrictions in the phase diagram. It can be raised up to 20 wt.% in the 'Internal Sn Diffusion' process which has a positive effect on the critical current density. Both processes are schematically presented in Figure 8.2.

Typical Nb_3Sn superconducting wire of 1 mm diameter consists of several thousand well-separated filaments of 3–5 μm diameter, which are embedded in a Cu-Sn bronze matrix (see Figure 8.2). Such filamentary configuration satisfies the requirements of electrical, thermal and mechanical stability.

In the past two decades there has been a steady increase in demand for superconducting high field magnets. The technology of high-field magnets is a rather specialized subject and we refer the reader to the literature cited at the end of

Figure 8.3: Photograph of hybrid magnet (courtesy of High Field Laboratory SNCI-CNRS/Max Planck–Grenoble).

this chapter. The highest field magnets usually have a separate core of the normal conductor and are referred to as hybrid magnets (see Figure 8.3). High field magnets are nowadays often used in medical applications, like Magnetic Resonance Imaging (MRI), for example.

8.2.3. Prospects

In the 1990's superconducting magnets with fields above 10 T are still expected to be constructed with wires based on A15 type superconductors. Other materials like Chevrel phases ($PbMo_6S_8$ or $SnMo_6S_8$) are still under development and, like HTSC oxides, may be of some importance for large scale applications beyond the year 2000. Fundamental limitations on the use of the high-T_c ceramics for magnet windings are related to the fact that these materials are quite brittle and they may not withstand the enormous stress exerted on the winding structure by high magnetic fields. This is already a limiting factor even with relatively ductile conventional superconductors.

Another fundamental limitation stems from the difficulty in passing a large current in a high-T_c superconductor. Flux pinning becomes an important problem. Firstly, one wants to operate at higher temperatures than in conventional materials and this favors fluctuations. Secondly, the energy involved in flux creep becomes smaller due to the smaller coherence length, ξ_0. As discussed in Chapter 5, ξ_0 decreases with T_c. Hence, one can predict difficulties in trying to pass high current through a room temperature superconductor, even if it were discovered!

8.3. Superconducting Electronics and Film Applications

8.3.1. Some applications of Josephson junctions

One of the best known applications of the Josephson junction is that of the voltage standard. It is based on the ac Josephson effect. As discussed in Chapter 6, the reaction of a junction to a microwave field is reflected in the I-V characteristics: sharp steps appear at multiples of a fixed voltage. The voltage V_n at the nth step is given by

$$V_n = n\Phi_0\nu_{\text{ext}} .\tag{8.2}$$

This relation between the dc voltage and the frequency ν_{ext} of the external signal depends only on fundamental constants. It is independent of the junction material or other properties of the junction. The frequency can be measured with very high precision so one can 'transfer' this precision to the voltage. The precision of the old voltage standard, the Weston cell of 1.018 V which had a stability of the order of ~ 1 ppm, is two to three orders of magnitude lower than the present standard based on the ac Josephson effect.

Another established application of the Josephson effect is in magnetometers based on the SQUID with either one (rf SQUID) or two junctions (dc SQUID). Both were discussed in Chapter 6. The field which is to be measured is usually coupled to the SQUID by means of a flux transformer.

SQUIDs require non-hysteretic junctions. Therefore one uses thin film microbridges or tunnel junctions shunted by a sufficiently small resistor. The sensitivity is mainly limited by the thermal noise of the junction resistance which depends on the square root of the signal bandwidth. Changes of magnetic flux of about $10^{-6} \; \Phi_0/\sqrt{\text{Hz}}$ can be detected, i.e., for flux through an area of $\sim 1 \; \text{cm}^2$ this corresponds to a magnetic field change $\sim 2 \times 10^{-17} \; \text{T}/\sqrt{\text{Hz}}$. Such high sensitivity has enabled the development of the evergrowing research in biomagnetism.

Josephson junctions also present a considerable potential interest for digital applications. The combination of switching speed of the order of picosecond and power consumption in the pW range provides impressive characteristics on which new electronic devices can be developed. In principle, one can construct logic and memory cells for fast computers. However, there are numerous other factors that have to be taken into account to evaluate this new technology, for example, the device reliability, very large scale integration, the very fast progress of semiconductor technology, economic profitability... . There are many other reasons that make it difficult for Josephson technology to even partially replace 'conventional' semiconductor technology; nevertheless it may gradually grow in several specific areas, metrology being one of them.

8.3.2. Conventional tunnel junctions

As we have discussed in Chapter 6, an important feature of Josephson tunnel junctions is their switching speed. It is fairly easy to give an estimate of the switching time of the junction: it depends on the thickness d of the tunnel barrier. For thinner oxide I_0 increases exponentially with decreasing thickness, whereas the capacitance increases only inversely with the thickness. Accordingly, the switching time decreases for thinner insulators. As an example, Pb-based junctions with a gap voltage of 2 mV, a specific capacitance of 5 μFcm^{-2}, and a current density of 2 kAcm^{-2} have a switching time of only 5 ps. Presently one can fabricate Nb-electrode junctions with high current densities that exhibit switching times of < 1 ps!

In commercial applications one uses mostly tunnel junctions of the S-I-S type for which there are already good fabrication techniques. Bridge type structures are also used, but the required small dimensions are still difficult to fabricate. For quick experiments, especially with oxide superconductors, point contacts are quite useful.

From fabrication point of view, the simplest device that exhibit Josephson effects is the *microbridge*. In this device the weak link is provided by a narrow constriction: as we have shown in Figure 6.7 there are several suitable geometrical configurations for a practical microbridge. Roughly speaking, the requirement is that the dimensions of the constriction, d, should be of the order of (or smaller than) the Ginzburg-Landau coherence length ξ:

$$d \gtrsim \xi . \tag{8.3}$$

At temperatures close to T_c these structures exhibit non-hysteretic I-V tunnelling characteristics as shown in Figure 6.14b. The hysteresis appears at lower temperatures, most likely due to isolated 'hot' spots which remain in the normal state due to the Joule heating.

Early experimental work on tunnelling junctions was done using soft, low melting point materials such as Pb, Sn or In, which are easily evaporated in vacuum. These junctions have shown I-V tunnelling characteristics closely corresponding to the predicted BCS behavior. However, these materials become considerably stressed when repeatedly recycled between 300 K and 4.2 K. This results in eventual one-dimensional overgrowths or whiskers that pierce the oxide layer causing shorts.

The best range of the required properties is to be found in refractory metal superconductors. They exhibit high mechanical strength, chemical stability, relatively high transition temperatures, and good adherence to the substrate, and form no protrusions or hillocks during thermal cycling. Most of the work on refractory metal junctions has been carried out with Nb deposited by electron beam evaporation or by sputtering.

Unfortunately even Nb technology is not without problems. Nb_2O_5 has a rather large dielectric constant ($\varepsilon \sim 36$) causing some limitations for its use as a tunnel barrier in devices for high frequency mixing applications. Moreover, dielectric oxide Nb_2O_{5-y} (where y is usually less than 1) is partially oxygen-deficient. Each oxygen vacancy is occupied by two Nb electrons that behave as donor sites. These localized states cause pinning of the Fermi level close to the conduction band. The oxide therefore behaves as an n-type semiconductor with a band gap of ~ 4 eV. These states act as sites inside the insulating barrier into which the quasiparticles may tunnel. This effect, known as resonant tunnelling, opens up additional tunnelling mechanisms at small bias potentials that cause increased sub-gap conductances.

Oxygen contamination of Nb decreases its transition temperature: ~ 1 at% of oxygen contamination in Nb lowers T_c by ~ 1 K. Further problems are also encountered when trying to make all-refractory tunnel junctions using Nb as the counter electrode. Any incoming Nb, as it is evaporated or sputtered over the barrier oxide layer, tends to getter (absorb) oxygen from the oxide tunnel barrier formed previously.

Josephson effects have also been observed in Nb devices in which the barrier is an oxide (for example; Nb-Al$_2$O$_3$-Nb), or a semiconductor (Nb-Se-Nb) or a normal metal (Nb-Ag-Nb). In the latter case measurable Josephson supercurrents were observed to flow over distances as large as 1000 Å. The weak coupling was achieved by the proximity effect in which there is a leakage of the order parameter into the adjoining non-superconducting material which contains some free carriers. Junctions made of Nb-Al$_2$O$_3$-Nb show a great deal of promise for electronic devices.

8.3.3. Preparation of tunnel junctions

We shall outline the preparation of conventional S-I-S tunnel junctions. One uses techniques which have some similarities with those used in semiconductor device technology.

Presently most commercial devices are made with Nb electrodes with Nb-oxide or Al-oxide tunnel barriers. Such junctions are very stable and can survive thousands of cycles between 300 K and 4 K. The Nb electrodes are deposited either by high-vacuum evaporation or by sputtering. The critical step is the oxidation of the Nb electrode and the formation of a thin Nb-oxide tunnel barrier. However, one can obtain very uniform and pin-hole free insulating barriers by using plasma oxidation.

The characteristic fabrication steps for preparation of an S-I-S junction are shown in Figure 8.4. First, photo-resist is applied to a substrate, which is then exposed to light through a suitable mask. The resist is removed from the exposed areas by the developer. Subsequently one deposits the metal electrode (by sputtering, for example). The resist (and the metal on top of it) are removed by chemical etching leaving the patterned base electrode. In the next processing step one produces a resist pattern with openings for the top electrode and carries out plasma oxidation of the oxide barrier. Finally, the top metal is deposited and by stripping off the resist one obtains the desired top electrode pattern of the completed junction. For industrial production one requires several additional processing steps like the deposition and patterning of the control lines over the junctions. Present technology enables one to fabricate several tens of thousands of working junctions on a chip.

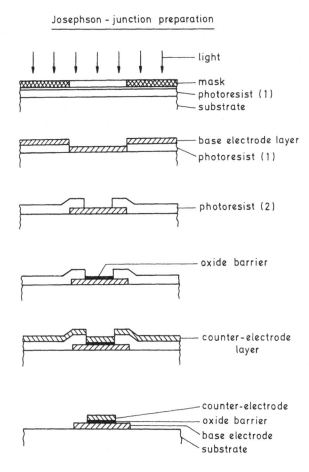

Figure 8.4: Schematic presentation of the processing steps required for fabrication of Pb-based tunnel junctions (after Wolf 1989).

8.4. Films of High-T_c Oxides

In this section we give a short, simplified overview of some of the preparation techniques and properties of thin films of high-temperature superconductors presently under active investigation. To readers interested in detailed technical aspects of the various deposition techniques, we recommend some of the specialized books and review articles listed at the end of the chapter.

8.4.1. Substrates

Before we describe the preparation of HTSC films, let us consider the most commonly used substrates.

The crystallography and preparation of the substrate is of primary importance in determining the quality of the deposited film. Some of the highest critical current densities ($J_c > 10^7$ Acm^{-2} at 77 K) and highest film uniformities have been achieved in YBCO films grown on commercially available (100) SrTiO$_3$ substrates. This is not surprising as the lattice parameter of this substrate differs only by $\sim 1\%$ from that of Y$_1$Ba$_2$Cu$_3$O$_{6.9}$. Moreover, the (100) direction ensures crystal growth with the c-axis oriented perpendicular to the substrate. By using (001) direction (and $\sim 100°$C lower substrate temperature) the c-axis is oriented in the plane; one therefore grows the a-axis oriented films which is of importance for tunnel junction applications. However, SrTiO$_3$ is fairly expensive, has high dielectric constant, and undergoes phase transition at ~ 110 K.

The LaGaO$_3$ substrate has lattice mismatch of only $\sim 0.5\%$ and exhibits a more favorable thermal expansion coefficient, a lower dielectric constant and more superior mechanical properties than SrTiO$_3$. However, it is neither cheap nor twin free. NdGaO$_3$ is reported to be lattice-matched $< 1\%$ and twin-free. It has low dielectric constant so it is probably the most promising choice of substrate for initial microwave type of applications. LaAlO$_3$ has been very successfully used, particularly for applications in passive microwave devices.

Many groups have successfully used cheaper (and less well-matched) MgO ($\sim 8\%$ mismatch) and yttrium stabilized zirconia (YSZ, lattice mismatch: $\sim 30\%$) substrates. It is clear that in the long run it will be necessary to use cheaper substrates like sapphire or silicon on sapphire (SOS) covered with suitable thin buffer layers like YSZ. Films can also be successfully deposited on silver layers, on sapphire substrates, or even on stainless steel substrates.

Most film depositions are carried out at substrate temperatures, $T_s > 650°$C, so one encounters problems with the interdiffusion between YBCO and the substrate. There is no universally accepted solution to this problem. To reduce the interdiffusion problems one would like to grow the film at the lowest possible substrate temperature, which is at present $\sim 600°$C.

8.4.2. Preparation techniques

In order to grow ~ 1000 Å thick superconducting YBCO film on (100) SrTiO$_3$ substrate (at $\sim 650°$C) one has to optimize, at least, the following processing parameters:

i) The stoichiometry, i.e., the correct "1:2:3" ratio between Y, Ba and Cu in $YBa_2Cu_3O_{6.9}$. One would like to control the stoichiometry within $\sim 1\%$, ideally within $\sim 0.1\%$ The former has been achieved; the latter will be more difficult to attain.

ii) Assuming that one has obtained the orthorhombic structure, it is necessary to ensure sufficient presence of oxygen in the film. For example, even if the metallic stoichiometry corresponds exactly to the 1:2:3 ratio, a film with only $O_{6.7}$ would exhibit zero resistance at ~ 60 K, rather than at ~ 90 K for the $O_{6.9}$ case. In particular, it is important to maintain the presence of oxygen in YBCO at the very interface with other layers within the planar device structure.

iii) To achieve critical current densities $J_c > 10^6$ Acm^{-2} at 77 K one has to achieve sufficient pinning in 'epitaxial' films. This has been readily achieved by now.

iv) For commercial device processing one requires uniform film deposition rate over wafer size dimensions. Ion beam and off-axis magnetron sputtering produce uniform films over several cm^2.

v) To demonstrate true Josephson tunnelling through a thin insulating barrier (~ 10 Å) one should master the epitaxial growth of atomically flat surfaces/interfaces. Despite the very rapid progress in this field such surfaces/interfaces are not readily grown and most tunnelling devices are effectively engineered microbridges.

There are many other parameters to be controlled and there are different solutions for different techniques. In what follows we briefly discuss ion beam sputtering and laser ablation. Other methods are discussed at length in specialized literature.

Ion beam sputtering

In Figure 8.5 we present a schematic diagram of the monotarget ion beam sputtering. It consists of a high vacuum chamber (base pressure 10^{-7} Torr) which contains an ion beam source, a water-cooled target holder and an oxygen inlet tube directed at the heated substrate holder. $Y_1Ba_2Cu_3O_{6.9}$ ceramics are used as the target. Films are deposited at $\sim 650°$C on (100) $SrTiO_3$ using the following typical sputtering parameters: Argon pressure 10^{-3} Torr, ion beam current 20 mA, beam voltage 500 V. Typical deposition rates are ~ 1 Ås^{-1}. After the deposition the chamber is filled with oxygen at a pressure of ~ 10 Torr and the substrate is cooled to ambient temperature with a 30' 'step' at 450°C. This procedure produces sharp resistive transitions ($\Delta T_c < 1$ K) with zero resistance at ~ 92 K and critical current densities above 10^6 Acm^{-2} at 77 K.

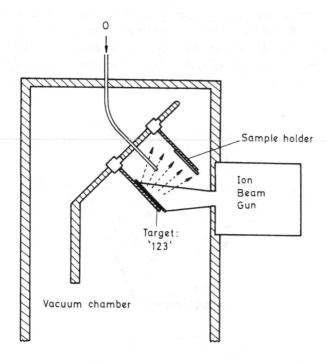

Figure 8.5: Schematic diagram of monotarget ion beam sputtering.

An important characteristic of ion beam sources is that they can also be used for ion beam etching. Ion beam etching, being anisotropic, is particularly well suited for etching fine line patterns such as those required for high-T_c SQUIDs. In addition to using photoresist as etch mask, metal overlayers patterned by plasma etching may also be used as mask during the ion beam etch stage. Such 'in-situ' ion beam etching capability is invaluable for pre-cleaning of substrates immediately prior to the YBCO film deposition.

Laser ablation

In principle the laser ablation is a fairly straightforward technique (see the schematic diagram Figure 8.6): the laser beam is focused onto a small ceramic $Y_1Ba_2Cu_3O_{6.9}$ pellet (typically ~ 1 cm in diameter) which is mounted opposite a suitable substrate holder. Using a laser beam power of $\sim$1–3 Jcm^{-2} the '123' material is directly 'evaporated' from the ceramic target onto the adjacent substrate:

The stoichiometry of the target is transferred into the film. It may be beneficial to use an oxygen flow directed onto the substrate, or preferably a reactive oxygen plasma. The films made by laser ablation produce, on average, the highest critical currents and one can use multiple target arrangement to grow YBCO-PrBCO superlattices.

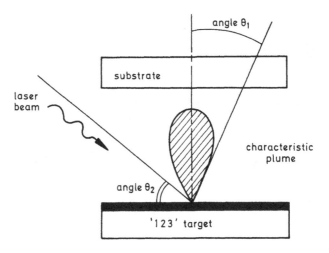

Figure 8.6: Schematic diagram of laser ablation technique.

The deposition rate is usually ~ 1 Ås^{-1} but even the rates of up to 100 Å^{-1} do not seem to be detrimental. The main disadvantage of laser ablation is that the film is deposited over a relatively small area and there is a tendency toward formation of droplets in the surface of the films. Such problems can be solved so this technique may turn out to be the most efficient synthesis route of HTSC oxide films.

Following the discovery of Bi- and Tl-superconducting oxides, several laboratories have successfully produced thin films by some of the above techniques and reported critical current densities of about 10^6 Acm^{-2} at ~ 100 K for some of these films. However, it is well known that thallium is a highly toxic element and great caution is necessary when one handles oxides of this element or cleans the evaporation system after several depositions.

Most Bi-films, including single crystals and ceramics, often contain at least two phases and the critical current appears to be very sensitive to external or local magnetic field. Contrary to YBCO reports on postprocessing, reports of patterning Bi-films are at present somewhat discouraging.

Finally, molecular beam epitaxy (MBE) is an experimental technique that may enable very fine control of the layer-by-layer growth (as in GaAs semiconductor research) of HTSC oxides, while the related technique, metal-organic chemical vapor deposition (MOCVD), might do it on the industrial scale.

8.4.3. Superlattices

We have shown in Table 7.2 that there is a series of compounds described by the formula $RBa_2Cu_3O_7$, where R represents one of the lanthanide elements that can replace yttrium in the original '123' structure of Figure 7.2. This means that '123' structure exhibits superconductivity with most of the lanthanides.

However, one of the exceptions is the $PrBa_2Cu_3O_{6.9}$ (PrBCO) compound which does not exhibit superconducting properties. This compound effectively behaves as a tunnel barrier in the temperature range where other compounds like YBCO exhibit superconductivity. PrBCO has a matching crystalline structure with YBCO. It can be grown with the a-axis perpendicular to the substrate between the two YBCO superconducting layers, resulting in an artificial, 'in-grown' tunnel barrier which is obviously of great potential interest for tunnelling devices. Such artificial structures may become very important for tunnel junction technology or for 'molecular engineering' of the new oxide structures.

8.5. Some Applications of HTSC Films

In this section we shall briefly outline some of the areas of research of interest on film applications of HTSC oxides. Most efforts are focused on three main applications areas:

i) High quality epitaxial thin films of low surface impedance for passive microwave components like resonators, antennas, etc.

ii) Advanced epitaxial heterostructures for novel semiconductor-superconductor hybrid devices or chip interconnects.

iii) Superconductor-insulator-superconductor (SIS) and superconductor-normal metal-superconductor (SNS) 'sandwich' structures for high-T_c tunnel junctions and weak link devices.

For some applications one has to be aware of the negative aspects of the device operation at liquid nitrogen temperature as compared to helium: there is evidently an increase in thermal fluctuations. This factor is especially important in devices which use Josephson junctions. In Chapter 6 we have discussed the temperature below which one has to operate as a function of the current of the junction. In various devices one uses a factor of 10 to 10^3 which means that at liquid nitrogen temperature the required value of I_c can be as large as 3 mA. This has two important consequences: it increases the power dissipated by the junction and requires very small value of the loop conductance; as we pointed out $\frac{2\pi L I_c}{\phi_0}$ has to be of the order of 3 in an operating device.

8.5.1. Passive HTSC microwave devices

This is the area where most progress has been made and where the first products are already marketed. HTSC stripline structures have been constructed for microwave use as filters, resonators, antennas and similar passive device structures. Many such devices are already being tested in satellites in outer space where the temperature of operation (80–100 K) is very suitable for HTSC films.

The requirement for such applications is good substrates with low microwave losses (like $LaAlO_3$), as well as high quality YBCO films with low surface resistance at high frequencies. The actual shape of the device is obtained by standard ('semiconductor') processing techniques, like photolithography, dry etching, etc.

An important property of HTSC films in this type of devices is their low surface impedance: for example, < 0.8 mΩ at 77 K in YBCO films at 10 GHz; this is almost two orders of magnitude lower than the surface impedance of copper. Therefore one can achieve higher performance of filters and resonators: Q up to 20000 at 77 K (at 10 GHz); up to two order of magnitude better than Cu or Au.

In Figure 8.7 we show an example of the preliminary design of HTSC antenna. Similar passive devices, resonators for example, are already commercially available from several companies.

8.5.2. Bolometers and fast optical detectors

HTSC bolometers take advantage of the steep change of temperature dependence of resistivity at T_c to achieve high responsitivity. The bolometric detection of radiation has many advantages, mainly due to to the essentially flat energy-responding detection characteristics and simple absolute calibration. The use of HTSCs is convenient in this type of application due to their low noise and high sensitivity. The theoretical possibility of realizing very low-noise bolometer for the far infrared range is potentially superior to any other existing detection method in the wavelength range $\lambda > 20$ μm.

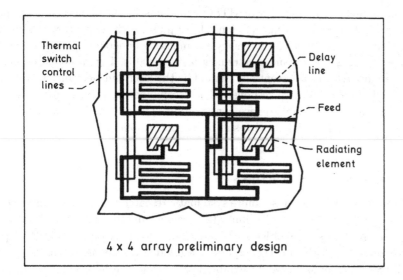

Figure 8.7: An example of HTSC antenna (after Romanofsky, NASA 1990).

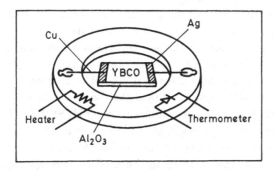

Figure 8.8: Schematic diagram of a HTSC bolometer (courtesy of Rowell, Conductus, 1991).

The detectivity of a bolometer realized on an HTSC film can reach ~ 800 V/W at 90 K and the noise equivalent power, NEP $\sim 5 \times 10^{-11}$ W/$\sqrt{\text{Hz}}$. Response time of 1 ms (and better) can be achieved. The performance of the bolometer can be further optimized by using processing techniques to increase the sensitive area and hence the output voltage of the detector. By selectively thinning the substrate, heat capacity and heat conduction can be further minimized. Several companies are developing the bolometric arrays to be used in pixel-arrays in imaging applications.

The non-thermal response of HTSC thin films in the structure of striplines can be exploited by use of fast (< 1 ns) laser pulses. Phenomena related to non-equilibrium superconductivity are not yet completely understood, but it seems that the pair-breaking mechanism is extremely fast: ~ 300 fs. Experiments have shown pulse response which was essentially limited by the laser and electronic speed rather than by the inherent device response. The recombination rate is rather slow, but can be improved by suitable device construction. The applications are for creating a fast detector and/or switch with sub-picosecond response.

8.5.3. Hybrid device structures

Hybrid device structures require systematic research on the use of intermediate buffer layers on technologically important semiconductors for (opto-)electronics, such as Si and GaAs. The objective is to evaluate the potential of HTSC striplines as interconnects. Although the experts are divided on this issue, one may argue that HTSC interconnects would exhibit lower capacitive losses and that, due to higher carrier mobility at 77 K, some novel hybrid applications may appear in the future.

YBCO films were successfully grown on Si with a variety of buffer layers; the most promising results were obtained with yttrium-stabilized zirconia (YSZ) and CaF_2 buffers. Several hybrid devices have been demonstrated on Si substrate. YBCO films were also grown on GaAs with ITO and MgO buffer layers. ITO is a 90%–10% mixture of indium and tin oxides that results in transparent conductive films that are used in photodetection and opto-electronic devices, in general. An optical modulator could be produced by using the electro-optic effect in GaAs and ITO films as electrodes. A similar modulator, using HTSC thin films in place of the ITO films, could achieve higher speeds (of the order of 10 GHz). There is a need of attaining high critical currents (of the order of 10^6 Acm^{-2}) for the modulator to perform at high frequencies, leading to the need to grow epitaxial films on GaAs (using suitable buffer layers, e.g. CaF_2 or MgO). Such attempts are in the very early stage of development and it will take several years to fully evaluate the feasibility of such devices.

8.5.4. Tunnel junctions and SQUIDs

This area of applications exploits the Josephson tunnelling effects and magnetic flux quantization which we discussed in Section 8.3.1 in the context of conventional superconductors. When trying to fabricate SNS and SIS devices from HTSC oxides, one faces several materials problems. The tunneling structures need very thin insulating layers, free from pinholes and defects, between two YBCO layers. Such structure is very difficult to fabricate due to the enhanced interdiffusion which occurs at the growth temperature ($\sim 650°C$) of YBCO. Due to the short coherence lengths of the HTSC materials, microbridge structures need to be very small (preferably < 100 Å), leading again to fabrication problems. The potentially higher operating temperatures of HTSC devices as compared with conventional weak link devices yield relatively higher noise levels. However, first SQUIDs made with HTSC junctions, which operate at 77 K, have been successfully demonstrated and may be of interest in geophysics and in other areas.

Various possibilities around the materials problems exist, e.g., using PrBCO (which has matching crystalline structure with YBCO) as insulating layer for the tunnel structure, and using Ag (which can have, by proximity effect, longer coherence length) for the SNS and microbridge structures. There is also the possibility to use natural barriers within the material, e.g. grain and twin boundaries.

Among HTSC oxides, the *isotropic* $BaKBiO_3$ ($T_c \sim 30$ K) exhibits BCS-like *I-V* tunneling characteristics and can be used as the electrode material in tunnel junctions which operate at 15 K. This is very useful as one can use closed cycle refrigerator rather than liquid He coolant. Other HTSC oxides, YBCO and particularly Tl-compounds, may eventually become of interest in some novel electronics applications at 77 K. The former is extensively used as a development test-HTSC material but most likely will not be extensively used at 77 K (remember the rule Eq. (8.1)), while the latter could be used at 77 K when its technology is fully developed. In particular, it is the potential toxicity of Tl-compounds, which requires special fabrication procedures, that aspect could prevent faster commercialization of this material.

8.6. Future Prospects

It is impossible to predict the evolution of the superconducting technologies but some of the trends are nevertheless visible. It is clear that in the foreseeable future the Josephson technology will not replace the dominating silicon technology or even represent serious competitor to the growing opto-electronic technology.

However, the number of applications of superconducting films will certainly grow, initially in passive microwave devices and portable SQUID magnetometers and later in some novel electronic components. The Nb and NbN based Josephson technology will dominate the field of high precision metrology (standards). The development of HTSC artificially layered structures will most likely have some (un)expected applications, not only in the in-grown tunnel junctions but perhaps in some novel three-dimensional electronic device-structures of the future. In other words, as the trend toward miniaturization in electronics is likely to continue, these 'doped' oxide superconductors might provide some unexpected applications within future electronics of sub-1000 Å structures.

So, while exact applications of HTSC films cannot be predicted, the range will probably be as large as for the conventional superconductors. The superconductor electronics includes the domains of detection (SQUID magnetometers, mixers, photon detectors, bolometers), interconnections (waveguides) and analog signal processing (Josephson junctions, A/D converters, superconducting FETs), as well as the domain of digital signal processing (memories).

In large scale technology, one should expect the continuation of the conventional Nb compound technology in wires, cables and magnets. HTSC oxides will most likely also gradually emerge in such applications, but most of them may become relevant early in the next century.

Regarding even higher T_c materials, the prospects are good from the physicist's point of view: we need an increase in critical temperature of only a factor of 3; we already have the right order of magnitude. However, materials scientists would argue that millions of man-hours have been spent in synthesizing new materials in the past few years, without appreciable success. Finally, engineers would argue that in any case we need T_c's of about ~ 500 K in order to design stable new devices which operate at room temperature.

Therefore there is no firm prediction or conclusion except that this is one of the most fascinating problems in contemporary physics. We can only hope that this textbook has given you an easy introduction into this stimulating field and that you, the reader, might eventually solve some of the numerous remaining puzzles which we discussed. And, who knows, it might be you yourself, 'the lucky one', who will some day demonstrate the challenging 'dream' of the superconductivity community — the room-temperature superconductor!

Summary

1. Most high current applications require
$$J_c > 10^5 \text{ Acm}^{-2} , \quad T_{use} \sim \frac{T_c}{2} ,$$
and for weak current and most applications in electronics
$$T_{use} \sim \frac{2}{3} T_c .$$

2. The most widely used materials for high field applications are Nb-Ti and Nb$_3$Sn. They are used at liquid helium temperatures and their upper critical fields are $B_{c2}(0) = 15$ T and 24 T respectively.

3. Refractory materials dominate conventional Josephson junction technology; Nb and NbN are the most widely used electrode materials of the 'liquid helium technology' that is very relevant to metrology.

4. HTSC oxides will have some impact in electronics-related applications: initially in passive microwave devices and portable SQUIDs. The mixtures of HTSC oxides with silver may offer somewhat improved properties and might provide the basis for future wires and cables.

Further Reading

A. Barone and G. Paterno: *Physics and Applications of the Josephson Effect*, John Wiley, New York, 1982

E. W. Collings: *Design and Fabrication of Conventional and Unconventional Superconductors*, Noyes Publ., Park Ridge, 1984

J. D. Doss: *Engineers' Guide to High-temperature Superconductivity*, John Wiley, New York, 1989

J. Evetts (editor): *Concise Encyclopedia of Magnetic and Superconductivity Materials*, Pergamon, Oxford, 1992

J. H. Hinken: *Superconductor Electronics*, Springer, New York, 1989

T. P. Orlando and K. A. Delin: *Foundations of Applied Superconductivity*, Addison-Wesley, 1991

S. T. Ruggiero and D. A. Rudman (editors): *Superconducting Devices*, Academic Press, London, 1990

T. van Duzer and C. W. Turner: *Principles of Superconducting Devices and Circuits*, North-Holland, Amsterdam, 1981

J. E. C. Williams: *Superconductivity and Its Applications*, Pion, London, 1970

Appendix I.   MAXWELL EQUATIONS

The most general equations are the following:

$$\text{curl } \mathbf{E} = -\frac{d\mathbf{B}}{dt}$$

$$\text{curl } \mathbf{H} = \mathbf{J} + \frac{d\mathbf{D}}{dt}$$

$$\text{div } \mathbf{D} = \rho$$

$$\text{div } \mathbf{B} = 0$$

$$\mathbf{B} = \text{curl } \mathbf{A}$$

$$\mathbf{E} = -\nabla V - \frac{\partial \mathbf{A}}{\partial t}$$

$\mathbf{E}$ = electric field in V/m
$\mathbf{D}$ = displacement field in C/m^2
$\mathbf{B}$ = magnetic field in T
$\mathbf{H}$ = excitation field in A/m
$\mathbf{J}$ = density of current in A/m^2
ρ = density of charge in C/m^3

In the vacuum, we have:

$$\mathbf{B} = \mu_0 \mathbf{H} \qquad \mu_0 = 4\pi \times 10^{-7}$$

$$\mathbf{D} = \varepsilon_0 \mathbf{E} \qquad \frac{1}{4\pi\varepsilon_0} = 9 \times 10^9$$

$$\varepsilon_0 \mu_0 = \frac{1}{c^2}$$

$$\sqrt{\frac{\mu_0}{\varepsilon_0}} = 376\Omega$$

Within a material we have:

$$\mathbf{B} = \mu_0(\mathbf{H} + \mathbf{M})$$
$$\mathbf{D} = \varepsilon_0\varepsilon\ \mathbf{E}$$
$$\mathbf{J} = \sigma\ \mathbf{E}$$

where we introduced the magnetization M in units of A/m, the dielectric constant ε and the conductivity σ in $(\Omega\mathrm{m})^{-1}$.

Appendix II.   FUNDAMENTAL CONSTANTS

elementary charge	e	1.6×10^{19} C
electron restmass	m	9.1×10^{-31} C
Planck constant	h	6.6×10^{-34} Js
	$\hbar$	1.05×10^{-34} Js
Bohr radius	$a_0 = 4\pi\varepsilon_0 \dfrac{\hbar^2}{me^2}$	0.5×10^{-10} m
Rydberg	$R = \dfrac{me^4}{32\pi^2\varepsilon_0{}^2\hbar^2}$	13.6 eV
Speed of light (in vacuum)	c	3×10^8 ms^{-1}
Permittivity of vacuum	ε_0	8.8×10^{-12} MKSA
Permeability of vacuum	μ_0	$4\pi \times 10^{-7}$ MKSA
Magnetic flux quantum	Φ_0	2×10^{-15} Wb
Josephson quotient	$2e/h$	4.8×10^{14} HzV^{-1}
Avogadro's constant	N_A	6.05×10^{23} mol^{-1}
Boltzmann's constant	k_B	1.38×10^{23} mol^{-1}
		8.6×10^{-5} eV K^{-1}
Bohr magneton	$\mu_B = \dfrac{e\hbar}{2m}$	9.3×10^{-24} JT^{-1}
electron volt	1 eV $\equiv$	1.6×10^{-19} J
	$\equiv$	2.4×10^{14} Hz
	$\equiv$	1.16×10^4 K
	$\equiv$	8.06×10^3 cm^{-1}
	$\equiv$	1.24 μm

QUESTIONS AND EXERCISES

Chapter 1

1.1. What are the two distinct properties that a material should exhibit in order to be considered a superconductor?

1.2. Try to understand why the magnetic field is not expelled within a sheath of depth λ at the surface of a superconductor.

1.3. Do pure metals with very low resistivity, like Au or Ag, exhibit superconductivity? Is there any magnetic element that is a superconductor?

1.4. Can conventional superconductors be used in a coolant other than liquid helium?

1.5. What are the main applications of superconductors?

Chapter 2

2.1. What is a Cooper pair? What is the density of pairs in conventional superconductor at zero temperature?

2.2. Using the values for critical temperature T_c and upper critical field B_{c2} given in Table 2.2 draw the phase diagram $B(T)$ for Nb. Assume that $B_{c1}(0) \approx 0.15$ Tesla.

2.3. Assuming that the coherence length ξ measures the length of the Cooper pair, estimate the number of Cooper pairs in a volume ξ^3 at $T = 0$. The result implies that the superconducting state cannot be considered as a gas of Cooper pairs.

2.4. In the mixed state of a type-II superconductor each vortex carries a flux $\Phi_0 = 2.067 \times 10^{-15}$ Weber. What is the number of vortices per m^2 in a material with an upper critical field $B_{c2} = 100$ Tesla?

2.5. What is the highest measured critical current density in a technologically important superconductor (like Nb-Ti) in zero field at 4.2 K? How does it compare with values for standard metallic wires?

2.6. Use Eq.(2.17) and Figure 2.8 to estimate the gap parameter Δ for Al, Nb and YBa$_2$Cu$_3$O$_7$. Their respective T_c's are given in Table 1.2.

2.7. Give the calculated gap parameters (from exercise 2.6) in terms of characteristic frequency (in Hz) and wavelength (in μm).

2.8. Use Eq.(2.20) and the data given in Table 2.1 to estimate the BCS coherence length of Al, Nb and YBa$_2$Cu$_3$O$_7$. Compare these estimates with the measured values given in Table 2.2.

2.9. Using the estimated values of the gap parameter of Al, Nb and YBa$_2$Cu$_3$O$_7$ (exercise 2.6) calculate the characteristic voltage drop V_t across a tunnel junction made of these superconductors.

2.10. What is the ac Josephson effect and what is the characteristic ratio ν/V (and λ/V) given in terms of fundamental constants? Use Appendix II to check the number given in the text.

Chapter 3

3.1. Using the values given in Table 2.1 and Eq.(3.10) calculate the approximate value of London penetration depth λ for Al.

3.2. Use Eq.(3.46) to calculate the upper critical field B_{c2} ($T = 0$) for selected superconductors listed in Table 2.2. Compare the calculated values with the experimentally determined ones given in the Table.

3.3. Use Eq.(3.58) to calculate the lower critical field $B_{c1}(0)$ of selected superconductors listed in Table 2.2.

3.4. Inserting values of B_{c2} (exercise 3.2) and $B_{c1}(0)$ (exercise 3.3) into Eq.(3.59) calculate the thermodynamic critical field $B_c(0)$ for the superconductors listed in Table 2.2.

3.5. Using Eq.(3.19) for the temperature dependence of previously estimated critical fields ($B_{c1}(0)$, $B_{c2}(0)$, $B_c(0)$) draw the $B(T)$ phase diagram for each superconductor listed in Table 2.2.

3.6. Using Eq.(3.8) demonstrate that the magnetic field inside a superconducting plate of thickness δ, perpendicular to the x-axis, is given by

$$h(x) = h_a \frac{\cosh(x/\lambda)}{\cosh(\delta/2\lambda)} \; ;$$

h_a is the external field parallel to the plate.

Chapter 4

4.1. Use Eqs.(4.16) and (4.17) to estimate the upper critical fields of anisotropic YBa$_2$Cu$_3$O$_7$ superconductor with $\xi^{ab} = 15$ Å and $\xi^c = 4$ Å.

4.2. Using values of ξ^{ab}, ξ^c, λ^{ab} and λ^c given in Table 7.4 for three different high-T_c oxides, calculate their critical fields, B_{c2}^{ab}, B_{c2}^c, B_{c1}^{ab} and B_{c1}^c as well as the effective mass ratio: m^c/m^{ab}; use the formulae given in Sec. 4.4.

4.3. Using Eq.(4.13) and the value for λ^{ab} calculated in the previous problem calculate the number of electrons n_s. Assume that m^{ab} is equal to the mass of a free electron.

4.4. Use Eq.(4.23) to calculate the depairing current for superconductors listed in Table 2.2. Use Eq.(3.58) to estimate B_{c1} and then Eq.(3.59) to calculate B_c. Compare the obtained values for depairing currents with measured values of the critical current densities given in Figure 7.11.

Chapter 5

5.1. Using Eq.(5.5) and the value of k_F given in Table 5.1 calculate the density of states at the Fermi level, $N(E_F)$ of Al.

5.2. Calculate the thermodynamic critical field B_c $(T = 0)$ for Al by using Eq.(5.57). Take the estimated density of states $N(E_F)$ of Al from the previous exercise and the value of T_c given in Table 2.2.

5.3. Derive the equation for the density of states of a superconductor in Eq.(5.62). Start with the dispersion relation Eq.(5.60) and assume that in the normal state the density of states $N(E)$ is constant, close to the Fermi energy.

5.4. Derive Eq.(5.51) by minimizing the total energy given in Eqs.(5.48), (5.49) and (5.50). Take into account Eq.(5.47) and the definition of the gap, Eq.(5.53).

5.5. Starting from Eq.(5.53) derive the self-consistent BCS equation (5.54) in the limit $T = 0$.

Chapter 6

6.1. Draw a schematic *I-V* diagram for an asymmetric SIS junction at zero temperature.

6.2. Check that Eq.(6.21) does indeed follow from Eqs. (6.9), (6.19) and (6.20).

6.3. Draw schematically the variation of the screening current vs. applied magnetic field in a dc SQUID.

6.4. Justify Eq.(6.75). Hint: Start from the Ohm's law, $\Delta V = R\Delta I$, and remember that the change in the critical current is twice the screening current.

6.5. Redraw Figure 6.19 in the case $\beta = 2\pi L I_0/\Phi_0 < 1$.

Chapter 7

7.1. Draw a schematic diagram of a unit cell of $YBa_2Cu_3O_6$ and $YBa_2Cu_3O_7$. What are the differences and where are the dopant-oxygen ions. Find the formal valence of a copper ion in a plane and in a chain.

7.2. Which elements can play a role of a dopant in $La_{2-x}M_xCuO_4$ and why?

7.3. What is the difference between $La_{2-x}Sr_xCuO_4$ and $Nd_{2-x}Ce_xCuO_4$?

7.4. Is there any isotropic HTSC compound?

7.5. Explain qualitatively why superconducting cuprates have large penetration depths and very short coherence lengths. Hint: see Table 2.1.

7.6. Compare the lattice parameter along the c-axis in Bi-2223 and YBCO 123 (see Figure 7.4). How large is the anisotropy of resistivity, i.e. ρ_{ab}/ρ_c ratio in these two compounds?

7.7. Compare the coherence length ξ_c and the lattice parameter along the c-axis in various high-T_c oxides. Do the same comparison for some of the conventional superconductors; what can you conclude?

7.8. What is the difference between the vortex lattice of a highly anisotropic oxide superconductor (like Bi-2223 compound, for example) and a conventional superconductor (like Nb)? What are the characteristic features of the mixed state of HTSC oxides?

7.9. What is the order of magnitude of the highest critical current densities (in zero field at 77 K) measured in epitaxial films and bulk ceramics; how much do these values get reduced in an external field of 1 Tesla? Why are critical currents of ceramic samples very low?

7.10. Use Eq.(7.5) to check whether the Ginzburg-Landau approach is valid in $La_{2-x}Sr_xCuO_4$ compound.

7.11. Estimate the characteristic temperature T^* for $YBa_2Cu_3O_7$ and compare it with the estimate for Bi-compound given in the section on Lawrence-Doniach model.

Chapter 8

8.1. What is a typical order of magnitude of critical current density (at 4.2 K in zero field) required for high current applications of type-II superconducting materials? Compare it with the depairing critical current of HTSC oxides (see Sec. 4.5.1).

8.2. Which Nb compounds are most often used in high field applications? What are their critical temperatures and upper critical fields?

8.3. Explain phenomenologically why thin filaments have to be used in order to prevent thermal runaway in high field magnets.

8.4. Which materials are most widely used in conventional Josephson tunnel junctions? What is the main technological challenge in the preparation of Josephson junctions?

8.5. Imagine a room-temperature superconductor; assume that v_F and B_c are of the same order of magnitude as in HTSC oxides. Discuss the problem of flux creep in such a material.

REFERENCES CONCERNING FIGURES AND TABLES

Bardeen, J.: *Physics Today*, AIP, 1990 [Fig. 2.7]

Buckel, W.: *Supraleitung*, Physik Verlag, Weinheim, 1977 [Figs. 4.1 and 6.7]

Coeure, J.: *Electronique* no. E124, Paris (1989) [Figs. 6.19 and 6.20]

Decroux, M. and Fischer, Ø.: *Electricité* **4**, Oct. 1988 [Fig. 2.4]

de Gennes, P. G.: *Superconductivity of Metals And Alloys*, W. A. Benjamin, New York, 1966 [Fig. 3.3]

Deutscher, G. and Müller, K.A.: *Phys. Rev. Lett.* **59**, 1745 (1987) [Fig. 3.5]

Doss, J. D.: *Engineers Guide to High-Temperature Superconductivity*, John Wiley, New York, 1989 [Fig. 4.7; Table 8.1]

Dugdale, J. S.: *The Electrical Properties of Metals and Alloys*, Edward Arnold, London, 1977 [Fig. 2.1]

Essmann, U. and Träuble, H.: *Phys. Lett.* **24A**, 526, 1967 [Fig. 4.3]

Evetts, J.: *Physics World*, Feb. 1990, p. 24, IOP, Bristol, 1990 [Fig. 7.11]

Flükiger, R.: *Supraconductivité* (ed. F. Lévy, P. Martinoli, C. Schüler, M. Q. Tran, and G. Vécsey), p. 201, 31st Proc. AVCP, Lausanne, 1989 [Figs. 8.1 and 8.2]

Friedmann, T. A., Rabin M. W., Giapintzakis, J., Rice, J. P., and Ginsberg, D. M.: *Phys. Rev.* **B42**, no. 10, p. 6217 (1990) [Fig. 7.7]

Hewatt, A. W.: courtesy of, (1991) [Fig. 7.4]

Jérome, D.: *Earlier and Recent Aspects of Superconductivity*, p. 113 (ed. J. G. Bednorz and K. A. Müller), Springer, Berlin, 1990 [Fig. 1.8]

Kim, Y. B., Hempstead, C. F. and Strnad, A. R.: *Phys. Rev.* **129**, 528 (1963) [Fig. 4.4]

Kittel, Ch.: *Introduction to Solid State Physics*, John Wiley, New York, 1986 [Table 1.1]

Livingston, J. D.: *Rev. Mod. Phys.* **36**, 54, (1964) [Fig. 4.2]

Lynton, E. A.: *Superconductivity*, Methuen, London, 1971 [Fig. 5.3]

Müller, K. A., Takashige M. and Bednorz J. G.: *Phys. Rev. Lett.* **58**, 1143 (1987) [Fig. 7.9]

Poole, C. P. Jr., Datta T. and Farach, H. A.: *Copper Oxide Superconductors*, John Wiley, New York, 1988 [Table 5.1]

Rose-Innes, A. C. and Rhoderick, E. H.: *Introduction to Superconductivity*, Pergamon, Oxford, 1969 [Fig. 6.13 and Table 6.1]

Romanofsky, R. R.: NASA/Lewis Center, Cleveland, Ohio (1990) [Fig. 8.7]

Rowell, J. M.: courtesy of Conductus Inc., Sunnyvale, California (1991) [Fig. 8.8]

Shukla A., Mieville L. and Affronte M.: EPFL – Diploma report 1989 [Fig. 7.8]

Strnad, A. R., Hempstead C. F., and Kim Y. B.: *Phys. Rev. Lett.* **13**, 794, 1964 [Fig. 4.4]

Tinkham, M.: *Introduction to Superconductivity*, McGraw-Hill, New York, 1975; Krieger, R. E.: Malabar, Florida, 1985 [Figs. 5.2 and 6.10]

Wolf, P.: *Supraconductivité* (ed. F. Lévy, P. Martinoli, C. Schüler, M. Q. Tran, and G. Vécsey), p. 201, 31st Proc. AVCP, Lausanne, 1989 [Figs. 6.11 and 8.4]; *Cryogenics* **18**, p. 478 (1978) [Fig. 6.17]

Wolsky, A. M., Giese R. F. and Daniels E. J.: *Scientific American*, p. 45, Feb. 1989 [Fig. 1.9]

Worthington, T. K., Holtzberg F. and Field C. A.: *Cryogenics* **30**, p. 417 (1990) [Fig. 7.10]

Yvon, K.: in *Superconductivity in Ternary Compounds Vol. I*, p. 87 (ed. Ø. Fischer and M. B. Maple), Springer, Berlin, 1982 [Fig. 1.6]

Acknowledgements

M. Cyrot would like to acknowledge J. C. Vallier for photograph 8.3, comments of D. Lollman and the hospitality of the Physics Department at EPFL. D. Pavuna would like to acknowledge numerous comments and help of M. Affronte, B. Dwir, P. Fivat, F. Frangi, A. Gauzzi, B. J. Kellett, J. H. James, stimulating discussions with N. Affolter, H. Gilgen, V. Gasparov, J. van der Maas, H. J. Scheel, J. L. Tholence, J. C. Villegier, and I. F. Schegolev. The friendly support of Ch. Gruber, M. Ilegems, J. L. Martin, E. Mooser, B. Vittoz and all colleagues in the Physics Department of EPFL is highly appreciated. Finally, we gratefully acknowledge Centre National de la Recherche Scientifique (CNRS) and Association Vaudoise des Chercheurs en Physique (AVCP).

INDEX

239